Girlhood and the Plastic Image

INTERFACES: STUDIES IN VISUAL CULTURE

Editors Mark J. Williams and Adrian W. B. Randolph, Dartmouth College

This series, sponsored by Dartmouth College Press, develops and promotes the study of visual culture from a variety of critical and methodological perspectives. Its impetus derives from the increasing importance of visual signs in everyday life, and from the rapid expansion of what are termed "new media." The broad cultural and social dynamics attendant to these developments present new challenges and opportunities across and within the disciplines. These have resulted in a transdisciplinary fascination with all things visual, from "high" to "low," and from esoteric to popular. This series brings together approaches to visual culture — broadly conceived—that assess these dynamics critically and that break new ground in understanding their effects and implications.

For a complete list of books that are available in the series, visit www.upne.com.

Heather Warren-Crow, *Girlhood and the Plastic Image*

Heidi Rae Cooley, *Finding Augusta: Habits of Mobility and Governance in the Digital Era*

renée c. hoogland, *A Violent Embrace: Art and Aesthetics after Representation*

Alessandra Raengo, *On the Sleeve of the Visual: Race as Face Value*

Frazer Ward, *No Innocent Bystanders: Performance Art and Audience*

Timothy Scott Barker, *Time and the Digital: Connecting Technology, Aesthetics, and a Process Philosophy of Time*

Bernd Herzogenrath, ed., *Travels in Intermedia[lity]: ReBlurring the Boundaries*

Monica E. McTighe, *Framed Spaces: Photography and Memory in Contemporary Installation Art*

Alison Trope, *Stardust Monuments: The Saving and Selling of Hollywood*

Nancy Anderson and Michael R. Dietrich, eds., *The Educated Eye: Visual Culture and Pedagogy in the Life Sciences*

Shannon Clute and Richard L. Edwards, *The Maltese Touch of Evil: Film Noir and Potential Criticism*

Steve F. Anderson, *Technologies of History: Visual Media and the Eccentricity of the Past*

Dorothée Brill, *Shock and the Senseless in Dada and Fluxus*

Janine Mileaf, Please Touch: *Dada and Surrealist Objects after the Readymade*

J. Hoberman, *Bridge of Light: Yiddish Film between Two Worlds*, updated and expanded edition

Erina Duganne, *The Self in Black and White: Race and Subjectivity in Postwar American Photography*

Eric Gordon, *The Urban Spectator: American Concept-Cities from Kodak to Google*

Girlhood and the Plastic Image

HEATHER WARREN-CROW

DARTMOUTH COLLEGE PRESS • HANOVER, NEW HAMPSHIRE

Dartmouth College Press
An imprint of University Press of New England
www.upne.com

Manufactured in the United States of America

Designed by April Leidig
Typeset in Garamond Premier Pro
by Copperline Book Services, Inc.

University Press of New England is a member of the
Green Press Initiative. The paper used in this book meets
their minimum requirement for recycled paper.

Hardcover ISBN 978-1-61168-573-2
Paperback ISBN 978-1-61168-574-9
Ebook ISBN 978-1-61168-575-6

Library of Congress Cataloging-in-Publication Data
available upon request.

5 4 3 2 1

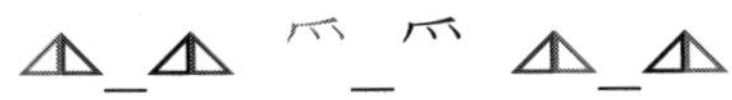

—EMILIE GERVAIS AND SARAH WEIS

contents

acknowledgments

This book is full of lists. Here is the first and the best:

My gratitude goes to my editor, Richard Pult; former professors Neil Cox, Shannon Jackson, Russell Merritt, Anne Nesbet, Kaja Silverman, Shannon Steen, and W. B. Worthen, my dissertation adviser; fellow Berkeleyans Patrick Anderson, Mona Bower, Heather Butler, Renu Cappelli, Gretchen Case, Scott Ferguson, Katie Gough, Kristina Hagström, Beth Hoffman, Laura Levin, Kelly Rafferty, Lara Shalson, Deborah Shamoon, Monica Stufft, and Joanne Stoddard Taylor; Center for 21st Century Studies denizens John Blum, Rachel Buff, Rebecca Dunham, Mary Mullen, Michael Newman, Rebekah Sheldon, Nathaniel Stern, Kristin Sziarto, and Charlotte Frost, the girliest girl I know; interlocutors Scott Bukatman and Richard Grusin; bold and generous reviewer Kirstin Ringelberg; artists Curt Cloninger, Charles Csuri, Emilie Gervais, Pierre Huyghe, Parker Ito, Martine Neddam, Philippe Parreno, Pedro Vélez, and Sarah Weis; Catherine Belloy at Marion Goodman Gallery, editor Valerie Valentine, without whom this book would be image free; editors Christi Stanforth and Amanda Dupuis; former student Donna Neal; Carol Edwards, Andrew Martin, and Brian Steele at Texas Tech University; my mom and dad; and Seth Warren-Crow, an ever-loving, ever-ready proponent of Girl Power.

Research for this book was supported by the Center for 21st Century Studies at the University of Wisconsin, Milwaukee.

preface

Lev Manovich's now canonical book *The Language of New Media* begins with a surprising revelation: "The avant-garde masterpiece *Man with a Movie Camera*, completed by Russian director Dziga Vertov in 1929, will serve as our guide to the language of new media."[1] Manovich's argument is that "cinematic ways of seeing the world . . . have become the basic means by which computer users access and interact with all cultural data."[2] New media are essentially (but not entirely) cinematic, he claims, and Vertov's work, in particular, prefigures the "database imagination" that would later produce digital objects from websites to 3-D CGI animations and computer games.[3] Moreover, *Man with a Movie Camera*'s double-exposures and other special effects — including its recursive opening shot of the man hauling his titular machine up a mound of dirt, all perched atop a gargantuan movie camera — anticipate the primacy of digital compositing through layer-driven software applications such as Photoshop and After Effects. While Manovich's reliance on cinema to establish much of the fundamental vocabulary of new media is critiqued by those concerned about the loss of specificity or troubled by the subsumption of film within narratives of new media's development, *The Language of New Media* is still an oft-cited primer on the formal properties of the digital. In fact, my own book makes use of his notion of "variability," one of five "principles" that Manovich offers to distinguish new and old media, despite their many continuities.[4] "A new media object," or, more particular to this book, a digital image, "is not something fixed once and for all, but something that can exist in multiple, potentially infinite versions."[5]

Although I did not set out to write a rejoinder to *The Language of New Media*, I now realize that *Girlhood and the Plastic Image* is, in one way, a kind of inadvertent but needy reimagining, conjuring a different avatar for an exploration of variability, multiplicity, potentiality, and infinity in digital image culture. What if our guide were not *Man with a Movie Camera*, but the character Alice in Wonderland? How does her transmedia nature as a figure visualized through photography, literature, illustration, philosophy, game design, and, yes, also cinema shift the terms through which we make sense of the digital condition? How can she expand Manovich's cinematic rubric? How might she redirect other pivotal concepts in media theory and visual culture studies toward the role of identity in the "cultural logic of media" (or, as I prefer, the cultural rhetoric)?[6] How is a girl falling through a rabbit hole different from a man ascending a mountainous camera?

While these questions are influenced by the technorati's enduring fascination with Carroll's novel, I turn slightly away from Marshall McLuhan's declaration that "Lewis Carroll greeted the electronic age of space-time with a cheer" and Ivan Sutherland's vision of a computer interface that "could literally be the Wonderland into which Alice walked."[7] In other words, *Girlhood and the Plastic Image* is moved not by the author or the world, but by Alice herself, her ability to use (and withstand) consumption as an activator of transformation, her capacity (and imperative) to change to suit different media and mediums, her stubborn engagement with problems of scale. Alice points us toward key attributes of digital images: malleability, transmediation, and instability. Furthermore — and this will become clear as the book proceeds — she encourages the formation of a congruent theory of plastic identity. The contemporary aesthetic subject, I argue, is a creature of variable scale and format modeled after the morphing, mobile, and girlish digital image.

This book is not about Alice in Wonderland, but about what she can help us understand. A minor can tell us something about digitality that a mister with a movie camera cannot.

Girlhood and the Plastic Image

one. (Un)Doing Girlishness

Plastic (Adjective)

1. Formative, creative <plastic forces in nature>
2. a. capable of being molded or modeled <plastic clay>
   b. capable of adapting to varying conditions: pliable <ecologically plastic animals>
3. sculptural
4. made or consisting of a plastic
5. capable of being deformed continuously and permanently in any direction without rupture
6. of, relating to, or involving plastic surgery
7. having a quality suggestive of mass-produced plastic goods; especially: artificial <plastic smiles>
8. relating to, characterized by, or exhibiting neural plasticity

Plastic (Noun)

1. a plastic substance; specifically: any of numerous organic synthetic or processed materials that are mostly thermoplastic or thermosetting polymers of high molecular weight and that can be made into objects, films, or filaments
2. credit cards used for payment—called also plastic money

—Merriam-Webster Dictionary, "Plastic"

i.

Plastic images preoccupy contemporary media culture. Our mediascapes are populated by elastic cartoon bodies, alterable avatars in online virtual worlds, and morphing "before" and "after" pictures on makeover shows. Magazines feature Photoshopped models, their thighs hollowed and bony edges digitally softened, and Frankensteinian celebrity mash-ups of older faces and younger bodies or vice versa. Software makes it easy to take a single image and export it in different formats, to extract a still from a moving image, to capture footage from a Massively Multiplayer Online Role Playing Game to make a narrative film, and to paste a mainstream actor's head on the body of a writhing porn star. Animated characters hawk products, play against us in video games, and express our emotions in text messages. And to date, nearly all of the top fifty highest-grossing films worldwide are animations or live-action movies emphasizing special effects.[1] What's special about these effects is actually what is commonplace: the image is plastic.

I have chosen the term "plastic image" for the multiple meanings of plasticity. Uncomfortably binding the aesthetic (the plastic arts) to the medical (plastic surgery), the economic (plastic money) to the material (plastic polymers), "plastic" operates reciprocally to designate either a force that molds or material that is molded. The plastic image conducts itself accordingly. Images are pliable; they can be sculpted like clay and circulated like money. Digital images in particular are known for their flexibility, for their acceptance of deformation without rupture. But images are not passive. They actively shape bodies and model identity. While the digital image might ask us to alter it, to upload and share it, to make it our own and then make it go viral, it also forges (and is forged by) a particular desire: the ability to make the embodied self adaptable to our needs, cross-platform compatible, and full of endless morphological potential. In Stephen Johnson's words, "Being digital means being able to reinvent yourself at the click of a mouse: morphing effortlessly from calculator to spreadsheet to word processor to video-editing console and back again."[2] This morphing is not always effortless (those who can't afford a computer, whose operating system is outdated, or whose bodies don't conform to the ergonomics of hardware might not tout the liberatory ease of their so-called digital selves), but it is integral to the performative rhetoric of digital images.

The unfixed image is a fundamental principle of digital culture. Commentators are in general accord. For example, while Lev Manovich's position on cinema may be controversial, his attention to the "variability" and "scalability" of digital images is shared.[3] Consider this particularly resonant statement from *The Language of New Media*: "In contrast to photographs, which remain fixed once they are printed, computer representation makes every image inherently mutable — creating signs that are no longer just mobile but also forever modifiable."[4] Although he and Manovich disagree as to the amount of divergence between analog and digital images, William J. Mitchell also identifies "the essential characteristic of digital information," and, by extension, the digital image, as the fact "that it can be manipulated easily and very rapidly by computer."[5] This openness to change affects our expectations regarding not only the speed and ease of modification, but also the scale at which change is understood. "Digital images are widely open to manipulation at a microscopic level" due to the discreteness of pixels and other image elements, notes Markos Hadjioannou.[6]

The discourse of digital plasticity has a long history. Since the advent of computer graphics in the 1960s, computers have been understood as transformers of images, able to shrink or enlarge an image, deform an initial image according to specific parameters, spin out numerous varieties of a source image, and, as early computer artist Frieder Nake tells us, "produce an entire class of drawings . . . running through a specific pattern in all its variations."[7] The curator and critic Jasia Reichardt highlights the agility and scalability of computer graphics: "The image on the cathode ray tube [screen] can be shifted, rotated, enlarged, seen in perspective, stored, recalled and transferred to paper with the intermediate stages recorded on film."[8] A reviewer of Reichardt's landmark 1968 exhibition *Cybernetic Serendipity* adds that "photographs, too, can be analyzed and stored in a computer's memory, then reorganized and distorted on electronic command."[9] And in a galvanic 1977 discussion of the capabilities of their prototype laptop, the Dynabook, Alan Kay and Adele Goldberg define the image generated by "drawing and painting" by computer as "a manipulable object" that "can be animated dynamically."[10]

Despite the appearance of geometric rigidity of most of these early computer images (especially when compared to their more fluid counterparts of the following decades), they were often understood as the products of potentially unpredictable processes of mutation and deformation. For example, Charles Csuri's "Sine Curve Man," winner of the 1967 *Computers and Automation* annual competition for computer art, uses an algorithm to transform an image of a face to give it a melting appearance (figure 1.1). Another well-known computer illustration is The Computer Technique Group's extraordinary but troubling "Running Cola is Africa" (1967/1968). They began with a series of three computer-generated line drawings — a man in a running pose, a Coke bottle, and the contour of the African continent — and

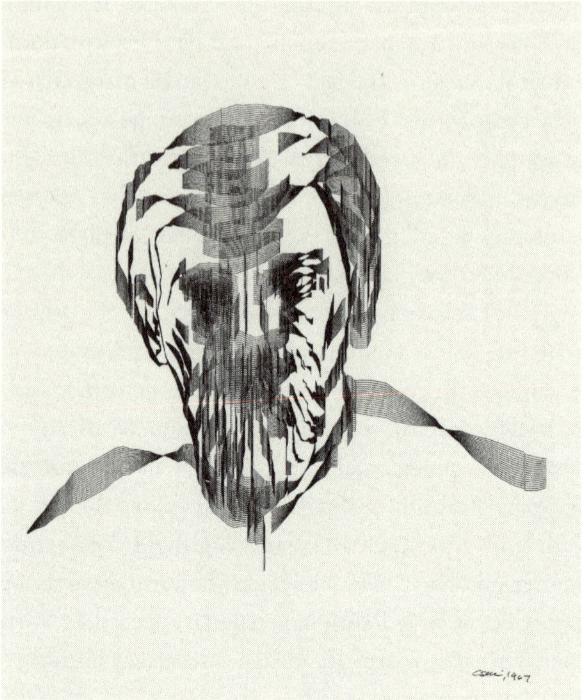

FIGURE 1.1
Charles Csuri, "Sine Curve Man," 1967. Courtesy of the artist

then digitally determined the multiple man-Coke and Coke-Africa hybrids between the keyframes, an interpolation technique for still images that "became the cornerstone of animation."[11] The rhetoric of mechanical reproduction is the making of sameness, even though an exact copy is impossible; computer generation promises replication, yes, but also, and more startlingly, an almost infinite variability and lability. Aptly, Nake deems the computer "a 'Universal Picture Generator' capable of creating every possible picture out of a combination of available picture elements and colors."[12]

The relative softness and alterability of computergraphic forms are elegantly presented in a 1972 demonstration of very early 3-D CGI animation technology.[13] While still a graduate student at the University of Utah, computer scientist Ed Catmull, who is now the president of both Walt Disney Animation Studios and Pixar, painstakingly digitized a plaster sculpture of his hand to produce an animated virtual model. He and fellow student Fred Parke shot a short film to document — or perhaps advertise — this new technique. The film gives us something like magic. The rigid plaster hand, once digitized and transformed into the wire-frame model on the computer, begins to relax; the fingers fan down one by one and open again in invitation. Then the hand is granted a more solid existence through the application of a crude, polygonal skin; finally, the rough edges are smoothed out and the hand adopts a rubbery appearance. After this sidelong reference to animation's hand-of-the-animator motif (what Donald Crafton calls the cartoon's performance of "parthenogenesis"), Parke shows off his own creation, a 3-D model of his wife's head produced using what we assume is the same process.[14] He doubles the image to create replicas that alternate between mirroring their expressions and moving their mouths independently, demonstrating both difference and sameness.

The generation of the hand and face is a process of ossifying the human form in realspace (the plaster model), then creating a more flexible, virtual form (the "wire" model), and finally giving it a "rubber" skin. Although the final products do not have the gestural suppleness of real bodies, the film insists on the flexibility of computer-generated images by showing the transmutation of plaster into virtual rubber. Indeed, this alchemical process is just as important to the film as its demonstration of the supposed realness of 3-D animation. The viewer is both invited to judge the animated models according to their verisimilitude and offered an unnatural perspective, quite literally: the virtual camera travels inside the animated hand, revealed to be hollow. Real hands have bones, of course, and plaster hands are thoroughly stiff, but CGI hands made of virtual wire and rubber have no internal structure. They have the potential for greater flexibility. A morphing title sequence created by Bob Ingebretsen performs this promise of 3-D CGI even before the process is explained: the production moniker "Halftone Animation" transforms into the letters of the animators' names. In fact, the credits' shape-shifting typography gives us images that are even more malleable than the 3-D hand, establishing a goal that can only be partially satisfied by available technology. The hand's shortcoming is less a failure, though, than an acknowledgment of the possibility and futurity of their technique. Evocatively, part of the hand sequence is later included in the science fiction feature *Futureworld* (1976).

The potential of mutable images is also expressed by one of the earliest examples of Internet theory. Four years before Catmull and Parke's demonstration of computer-assisted alchemy, computer scientists J. C. R. Licklider and Robert Taylor explained their concept of "living information" through the idea of malleable images.[15] Their essay "The Computer as a Communication Device" from 1968 presciently argues for a shift in thinking about computers as not senders and receivers of code but facilitators of interactive and collaborative communication, which they define in the following way: true communication occurs when two people who begin with different "mental models" come together through a reconfiguration of one or both ideas.[16] The cartoon illustrating this phenomenon shows "before" and "after" images (figure 1.2). In the first, two men in suits each think of an image that appears in comic speech bubbles hovering above their heads. The man on the left comes up with a voluptuous woman in a bikini; the one on the right visualizes a bicycle. In the extraordinary second picture, the man on the left has changed his mental image into a grotesque hybrid creature combining the girl and the bicycle, and the man on the right imagines his bicycle wearing the bikini. Thus, we learn, the success of their communication is proven by the morphing together of the two mental images. In Licklider and Taylor's words, "Creative, interactive communication requires a plastic or moldable medium that can be modeled, a dynamic medium in which premises will flow into consequences, and above all a common medium that can be contributed to and experimented with by all. Such a medium is at hand — the programmed digital computer."[17] In this example, the computer, the men's creativity, and Licklider and Taylor's vision of a proto-Internet fostering interactive "on-line communities" are all figured as shared plastic images.[18] While the cartoon is clearly intended as a symbol of communication and not a literal depiction of images online, it does register the importance of malleability to our early thinking about "living information" (of which Catmull's animated hand is another example) and presages a future in which the Internet allows people to communicate through shared images that transform. Further, according to the fractal logic of digital culture, the plastic image is simultaneously a thought, the computer that records that thought and allows it to be seen, and the proto-Internet that disseminates it and permits its modification. Plasticity itself is scalable and mobile.

The ontology of the digital image comes into clear focus for scholars of visual culture with the production and availability of home computers in the late 1970s and 1980s, improvements in Graphical User Interfaces (GUIs) in the 1980s and 1990s, and development of web browsers in the 1990s. By the

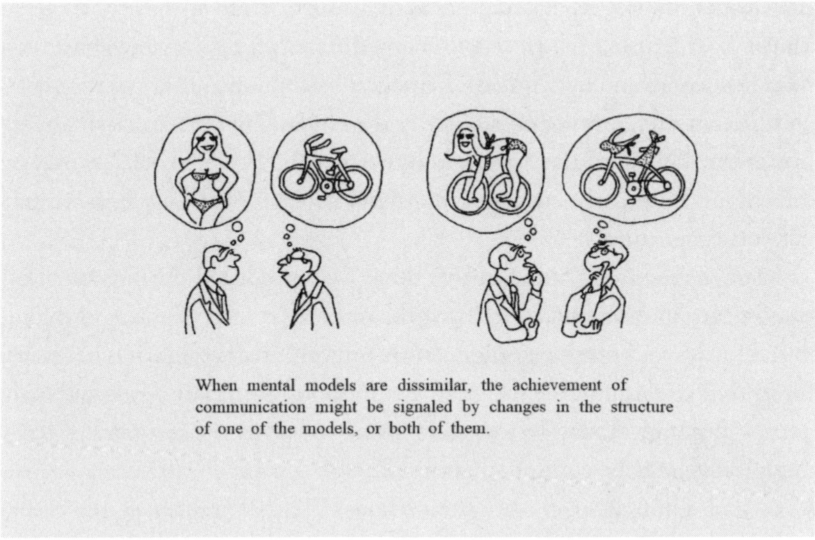

FIGURE 1.2 Illustration from J. C. R. Licklider and Robert Taylor, "The Computer as a Communication Device," reprinted from *Science and Technology*, April 1968. By permission of the Systems Research Center of Digital Equipment Corporation in Palo Alto, California

beginning of the millennium, the malleability of digital images is something of a truism, and by now it has become an observation so commonplace as to allow us to forget its impact on image consumption and production. Indeed, most scholarship on digital mutability comes out of a turn-of-the-century historical moment. Research on the digital imaginary after the early 2000s shifts from a focus on the variability of digital objects, the multiplicity of online personas, and the effects of digitization on the ontology of the photographic image to an emphasis on participation, sociality, real-time interaction, and Web 2.0. Despite this scholarly sea change, digital mutability hasn't lost momentum as a driving force of digital culture.

Web 2.0 expands our ability to resculpt and share images, while, in turn, the plasticity of data bodies facilitates large-scale participation in net culture. Writing in 2008, Ron Burnett explicitly connects the flexibility of the digital image to its ability to be shared in his discussion of the imograph, a neologism that designates "images that can be transformed through the use of software." Burnett continues: "Imographs also refer to the power of morphing and the ability to introduce a high degree of elasticity into digital images. Photographs and images of all sorts, from still to moving, no longer exist in isolation from the many different ways in which they can be transformed and networked."[19] Imographs embody our hopes for never-ending elasticity, infinite possibility, and a kind of facile connectivity made possible by high-speed networks. Manovich's principle of variability may be quoted less frequently, but it is a foundational tenet of twenty-first-century networked cultures.

Many of the case studies in this book were produced during a turn-of-the-century moment that witnessed the often-celebratory heyday of digital mutability as a scholarly paradigm. More importantly, though, it is necessary for us to revisit and refine the given assumptions of the late 1990s and early 2000s, although it may be tempting for us to match the fast pace of technological change by continually searching for new keywords. When Oliver Grau and Thomas Veigl state the obvious, "The digital image represents endless options for manipulation," as they did in 2011, we must respond with a close examination of endlessness and manipulability as the promises of the digital image.[20]

The central argument of this book is not only that these are the promises and premises of the digital. More pressingly, I claim that these attributes are not gender neutral. While other aspects of image culture are rigorously interrogated in relation to identity—take gaze theory's exegesis of classical Hollywood cinema, for example, or the feminist critique of antitheatrical prejudice—digital scalability and lability are not given the same kind of attention. What if we were to become sensitive to the dynamics of power inherent to the axiomatic attributes and functions of the digital image? What modes of identification are facilitated or hindered by endlessness and manipulability?

My proposition, advanced throughout this book's four chapters, is that our operative notions of image plasticity are entwined with conceptions of the plasticity of girls. In other words, Manovich's principle of variability is an embrace of girlishness—with all the disturbing and enabling implications of the word "embrace." If the objectification and to-be-looked-at-ness of women drives much of classical Hollywood cinematography, then the manipulability and to-be-shared-ness of girls powers postphotography. To return to what may have seemed like an oversight in my explanation of the data-driven conversation between those two men in suits: the liveliness and malleability of the computer, the communication it makes possible, and the data it sends into the world are explicitly presented as the girlification of a machine (the bicycle sporting a bikini), the mobilization of the girl (beach babe on wheels), and the yielding of both images to each other to generate creative variability.

Following Catherine Driscoll, a scholar who has been instrumental in the development of the field of girl studies, I work with an understanding of girlhood as "an idea of mobility preceding the fixity of womanhood and implying an unfinished process of personal development."[21] This is the cultural work that girls are expected to do. Their discursive duty is performed by their blossoming physiques, unstable body images, and variable identities — all of which are generated and exploited by an image culture obsessed with youth and transformation. Here I'm not only talking about images *of* girls. Additionally and more incisively, I'm claiming that some images *are* girls. Take, for example, music videos created by YouTube user and superfan Bunny Greenhouse (Chris Beckman) for songs by Girl Talk, the pseudonym of popular mash-up musician Gregg Gillis.[22] Bunny Greenhouse, a highly sophisticated amateur, produced videos that mimic the formal structure of the songs, which themselves are cleverly manipulated remixes of popular tracks. I propose that Girl Talk is not simply a name, but also a mode of engagement between consumers/producers and the plastic images and sounds that circulate across global networks. The mash-up — with the softness and mutability the word implies — exemplifies our contemporary obsession with the girlification of media objects. Our situation is one in which the reproduction, dissemination, and consumption of plastic images are viewed in conjunction with girlhood, and the figure of the girl is performatively produced as a soft(ware) image — a soft-copy, pliable, diminutive, *microsoft*.

Although images need not depict girls in order to be girlish, plastic images often do take girls and other juvenilized beings as their subjects. Think of the large number of young women, children, and cute animals in cartoons and the close relationship between women's fashion magazines and airbrushing. Also, consider "The First Lady of the Internet," Lena Soderberg, the twenty-one-year-old Swedish model who posed for what would later become the most common test image for algorithmic compression and denoising studies.[23] A year after Parke turned his wife's head into a 3-D animated model, electrical engineers at the University of Southern California's Signal and

Image Processing Institute scanned a page from *Playboy* magazine to use in their research. The result is a 512-by-512-pixel image of the centerfold, slightly lengthened due to problems with analog to digital conversion and cropped at the shoulder. Looking coyly at the camera while standing in front of a mirror, Soderberg wears a feathered hat at a jaunty angle. In most versions of the image we see a slice of her reflection and, behind her, some wooden beams. The full-length picture is more troubling; it reveals that she is in an attic playing a kittenish game of dress-up — or undress-up, as she is wearing only accessories — for an audience of toys, including a baby doll that reaches for her nude buttocks. Soderberg, also called an "Information Age Madonna" (although, as I argue later in this book, Madonna is the Information Age Madonna), lends her name to the digitized image, which is known simply as "Lena" or "Lenna."[24] According to Jeff Seideman, the former president of the Society for Imaging Science and Technology, "The use of her photo is clearly one of the most important events in the history of electronic imaging."[25] Writer Brian Merchant, not an engineer and thus without hands-on knowledge of the image, notes that he is familiar with Lena even though he can't remember where he saw her. "I'd like to think the image is just so deeply ingrained into internet DNA that Lena evokes some kind of digital déjà vu," he suggests, "but it's probably just because she's gorgeous."[26] The Bambi-eyed model, whose face has been subjected to every kind of image manipulation imaginable in a digital wonderland, recently made headlines when researchers in Singapore shrunk an already-cropped Lena down to a hair's width of 50 square micrometers (plate 1). Lena is the ultimate plastic image — not only microsoft, but also nanosoft, and Forever 21.[27]

The rest of this chapter will situate the plastic image within digital and (as will later become clear) protodigital discourses of adolescence and mediation.[28] First, I will address modern and late-modern conceptions of the malleability and elasticity of youth. Next, I will argue that Alice in Wonderland is a protodigital figure who emerges at the intersection of notions of girlish transformation and certain filmic practices (the production of photographic series, special effects cinema, and animation). Alice's storied history can help us see the development of the plastic image as a promise and problem of scale. Then, I will move from media practice to media theory, presenting its various approaches to the plastic image as a struggle between girlphobia and girlphilia (as well as, more broadly, infantophobia and infantophilia). In such fields as visual culture studies and media studies, the immature subject has long been at the center of debates concerning the agency of the viewer. Critiques of mass and consumer culture, in particular, are frequently made at the expense of the girl, whose supposed impressionability, instability, and openness to commodification are understood as indexes of and contributors to capitalist hegemony. All too often, such adultist arguments can't think of disempowerment and resistance outside of the opposing forces of girlification and maturation. However, as Alice and her fellow plastics can attest, girlphilia in media discourse does not usually lead to improvements in the lives of girls, and the labile, mobile, transmutational subject has been coopted by the Girl Talk of neoliberalism and globalization.

The plastic image is riven by such irreconcilable differences. Bearing the attributes of openness (or vulnerability), connectivity (or dependence), variability (or instability), and flexibility (or weakness), it has come to stand for what digital media can or can't offer us and what forms of selfhood are, aren't, or should be available. What is often thought to be peripheral, silly, and minor is actually at the heart of media cultures' most fervent investments in the power and failure of image making. The most pressing question isn't whether girlish images peddle nothingness or no-thingness, whether they are really the empty faces of American cultural imperialism or figures of resistance to objectification. More fundamental is how nothingness and its corollaries, openness and potentiality, have become attached to girls and images in the first place.

Carroll's novel describes getting bigger as "opening out like the largest telescope that ever was" (figure 1.3).[29] What does this telescope allow us to see?

ii.

"But I'm not a serpent, I tell you!" said Alice. "I'm a — I'm a — "

"Well! What are you?" said the Pigeon. "I can see you're trying to invent something!"

"I — I'm a little girl," said Alice, rather doubtfully, as she remembered the number of changes she had gone through, that day.

— Lewis Carroll, *Alice's Adventures in Wonderland*[30]

When consumers become producers, one of the objectives of their creative activity is the construction and multiplication of new selves.
—Grant McCracken, *Transformations: Identity Construction in Contemporary Culture*[31]

"Girl" is a capacious word. Female children are, of course, girls. Working women such as maids, assistants of various types, boutique employees, hair stylists, and sex workers are also often called "girls," sometimes attached to a belittling possessive pronoun. Alternately, "girl" can be an affectionate term between adult women, suggesting a kind of intimacy and candor; this sense of the word was used to promote the first *Sex and the City* movie (2008), which bore the tagline "The girls are back!" Applicable to the girls from *Sex and the City* is the American slang phrase popular on cable makeover shows: the imperative to buy a supportive bra and "show off your girls."

In regard to these four women, "girl" reminds us of their perpetually youthful bodies reshaped by surgery and dieting as much as it indicates our assumed familiarity with them. Cosmetic surgery is a practice sutured, so to speak, to popular understandings of girlishness. Indeed, as Kathleen Sweeney states, "'plastic' surgery suggests the production of a more bendable, moldable, modifiable self, in many ways just like Barbie."[32] Cosmetic surgery is a performative enactment of the malleability of the body. It girls the flesh, asserting that girlhood is less a state of being than a state of becoming, a way to assert and maintain the plasticity of an image and, by extension, a self. It is not only a means to achieve a goal, but also—and more importantly—a way of keeping the ball in play.

The construction of the self as a modifiable image must be thought in relation to the plasticity of the image in digital culture. Manovich's claim that computer representation produces "signs that are no longer just mobile but also forever modifiable" could just as well be part of a conversation about plastic surgery or its less invasive relative, Photoshop.[33] Case in point is a new mother who receives a makeover on a cable television program. Looking at a 3-D CGI image of her body produced by the show, she complains about the changes in her figure due to pregnancy. "I need a trainer or some surgery," she says, frowning at the avatar in front of her. Then, she states with more resolve, "It's a picture. It's an image. Everything can be changed."[34] Here the mother's desire to get her girlish figure back is one of turning her body into an image with an ontology based on computergraphic transformation. Girl-

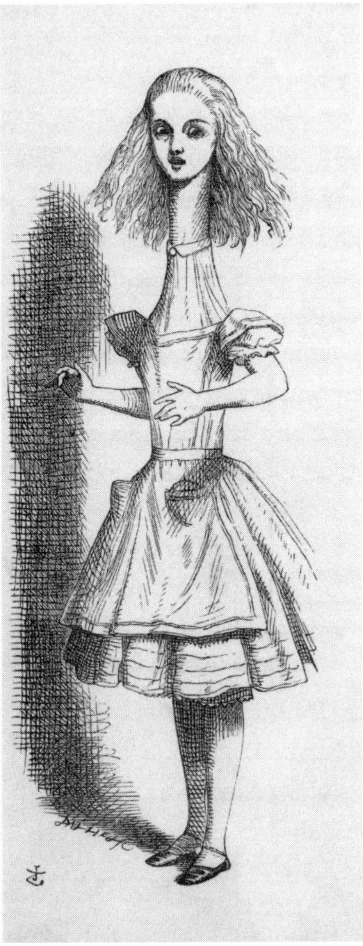

FIGURE 1.3
Alice grows taller, from *Alice's Adventures in Wonderland* by Lewis Carroll, published 1891 (litho). John Tenniel (1820–1914). Private collection / Ken Walsh / Bridgeman Art Library

ishness is thus the pursuance of a mobile and forever modifiable body. The 3-D representation of her figure is not really a representation, after all; it is the body that she will later try and fail and try again to produce by resculpting her flesh through diet, support garments, or surgery. In this future bodied forth by the image, the woman will be older, of course, but she will also be younger through the ongoing process of girlification. The plastic image promises a future that is both a nostalgic reclamation of the past and an unending but ever-changing present.

In key ways, the temporal and spatial elasticity of girlhood is an extension of attitudes forged in the early twentieth century, once adolescence emerged as a psychological, sociological, educational, and economic category. Robert Latham argues that Americans in particular have inherited their conceptions of young people from the 1920s and 1930s, when "youth enjoyed a special privilege within industries undergoing Taylorization, since this radical restructuring of production mandated an energetic pliancy in workers."[35] We continue to perform their "surreptitious conflation of 'youth' as a commonsense natural category . . . with 'youth' as a phantasmatic, fecundating energy potentially animating all persons regardless of chronological age."[36] Latham finds the late twentieth-century expression of the animation, adaptability, and malleability of the young-at-heart in such diverse sites as vampire movies and discourse surrounding Silicon Valley start-ups. Sweeping changes in production — Fordism and Taylorism on the one hand and postindustrialization on the other — require malleable and nimble worker-citizens; the consumption of products imbued with youthful and lively plasticity is imagined to transfer such qualities to voracious subjects. Interestingly, Thomas Lamarre notes that Studio Ghibli's Hayao Miyazaki and his mentor Yasuo Ōtsuka like to hire animators who are between eighteen and twenty-five years old because they are believed "to impart a sense of energy, vibrancy, and flexibility to the final product" (and, I would add, to confer such attributes to the viewer). "The emphasis is not on labor hours or quantity but on quality, and the quality of youthful energy is especially hard to define. It is rather like the wind."[37]

Although Latham doesn't advance an argument related to girlhood specifically, his characterization of consumption as the primary conduit of juvenilization invokes the long-standing relationship between femininity and commodification. Women, especially young women, are believed to excel at consumption and to welcome the commodity's external influence. Consider the contributions to the edited volume *The Modern Girl around the World: Consumption, Modernity, and Globalization*.[38] They present the Modern Girl of the 1920s and 1930s, a globalized phenomenon nourished by advertising campaigns for clothing and makeup, as not only a remarkably resilient image of commodity consumption able to adapt to numerous national contexts, but also itself a kind of phage, incorporating local ideals of feminine beauty within its contours. The Modern Girl was defined by an ability to consume; this faculty both incurred accusations of fatuity and facilitated her participation in the public sphere. As Latham's vigorous, flexible youths were driving modern production in America, Modern Girls in India, Shanghai, Germany, South Africa, and, yes, America (among other places) were using factory-produced commodities to refashion themselves and attain some degree of social flexibility. And as Latham portrays the vivacious youthfulness of the ideal American factory worker as an alchemical power enlivening products as well as consumers "regardless of chronological age," *The Modern Girl around the World* sees "the historical emergence of 'girl' as a modern social and representational category and as a style of self-expression [as] largely delinked from biological age."[39]

Sociologist Anita Harris, like Latham and the researchers of the Modern Girl, emphasizes the role of young people in performing modernity. However, Harris begins her account of what she calls "the ideal of the future girl" with adolescents living several decades earlier.[40] Noting the similarities between fin de siècle youth and twenty-first-century girls, Harris argues that "girlhood operates now as adolescence functioned then": now, it's "young women [who] are supposed to embody the new flexible subjectivity."[41] The normative girl of today is a paradigmatic citizen, employee, and subject, "self-making, resilient and flexible"—with the responsibility to adapt that was thrust upon late nineteenth-century adolescents and the transformative energy of Latham's early twentieth-century factory workers (if not the resistant spirit of some of the Modern Girls).[42]

While Harris claims that twenty-first-century girls are the adolescents of yore, Catherine Driscoll insists that adolescence has long been girled. In the work of psychologist G. Stanley Hall, often called the father of adolescence, the period between young childhood and adulthood is "characterized by feminized attributes such as changeability and malleability even when the adolescents are men."[43] Although Hall neglects to give girlhood its due attention, favoring instead to focus on a generic adolescence implicitly understood as boyhood, he "incorporates this feminization as a positive aspect of adolescence even while marginalizing and dismissing girls."[44] However, he does explicitly address girlhood in "The Budding Girl," a chapter in volume 2 of the earnest but sometimes ludicrous *Educational Problems*. The girl, he bemoans, "is the most intricate and baffling problem perhaps that science has ever yet attacked."[45] He nonetheless offers an analysis of her nature ("wild with a charming, gamey flavor"), eating habits ("too excited to feed well on plain, nutritive viands"), faddishness, likeliness of becoming hysterical under the pressures of urban modernity and propensity for "moulting old ideas and sprouting new ones of very different hue and composition," of her brain that hasn't hardened and her tendency to "act collectively and [be] limp and inert alone."[46] Most important to my argument is his assertion that "everything within and without tends to keep her plastic and she must remain adaptable to a wide range of eventualities."[47] Counterintuitively, Hall both sees the girl's vulnerability as a feminine pathology and claims girlhood as a supposedly pure and definitive expression of humanity.

Whether girlhood in late modernity has expanded to embrace attributes that used to belong to youth more generally or whether adolescence has always been feminized, plastic girlishness is less an expression of chronological age than a combination of feminization and infantilization that can be applied to men, women, children, and images. In fact, feminine adolescence is both an aesthetic form and what Thomas Streeter (following John Frow) calls an "imaginary for[m] of selfhood."[48] Streeter asserts that forms of selfhood are "not types of individuals. They are discursive patterns embedded in institutions and historical processes that become available to individuals as ways of making sense of who they are in given contexts."[49] Streeter continues, explaining the mobility and changeability that define this model of identity:

> "Imaginary forms of selfhood," then, are neither fixed identities nor complete or determinate in some kind of mechanical way. They are plural and fluid, but not infinitely so; there are typically several forms available to any given individual in any given context, and it is possible, and probably sometimes necessary, to move among them.[50]

Girlhood is thus not only an imaginary form of selfhood, but also a metaform of selfhood. The fluidity, instability, and variability that characterize imaginary forms of selfhood when viewed collectively are held within feminine adolescence's sprawling boundaries (contours designed to appear more infinite than Streeter imagines).[51] If forms of selfhood are "discursive patterns . . . that become available to individuals," then metaforms are the patterns of availability itself.

As I argue throughout this book, digital images mobilize feminine adolescence as a metaform of selfhood that produces the girlishness of computer-generated possibilities and the possibilities of girlishness. Consider an important but undertheorized moment in the digital imaginary: Nicholas Negroponte's gendered definition of multimedia from 1993. In an editorial entitled "Repurposing the Material Girl," Negroponte begins by noting that Madonna, a then "34-year-old former Michigan cheerleader," was signed by Time Warner for a "$60 million 'multimedia' contract."[52] He uses Madonna's multiplatform and metaformic persona as a launching pad for his discussion of the definition of multimedia. The Material Girl emerges in this thought piece as the best example of "the expanded use of text, audio, and video for as many purposes as possible, in multiple packages and through diverse channels"— not unlike the subject who would later arise from social networking websites (although Madonna leverages her multimedia identity for considerable personal gain, while the everyday social networker labors for free).

Madonna is an ideal embodiment of the plastic image. She stakes her career on her ability to re-create herself. She maintains that career through a continuing relevance to the youth market, a relevance produced, in no small part, by a preternaturally youthful physique morphed by plastic surgery and vigorous exercise. The Material Girl's own interest in the girling of the image can be read into her financial support of the 1997 Cindy Sherman retrospective at New York's Museum of Modern Art. Sherman is known for her "seemingly endless supply of chameleon identities" in such series as *Untitled Film Stills*, which the art historian Arthur Danto sees as a performance of variations on the figure of "The Girl."[53] Danto understands "The Girl" as "an allegory for something deeper and darker, in the mythic unconscious of everyone, regardless of sex."[54] While he has been taken to task for this statement — for example, Abigail Solomon-Godeau observes that Sherman performs older roles as well — I think that he is onto something, albeit unwittingly.[55] Sherman's performance of girlhood is independent of the age of her characters. "The Girl" is enacted through the plasticity, incompleteness, and accessibility of her personas; she can be "the mythic unconscious of everybody" because she is perceived as a nobody. Likewise, Madonna, as the emissary and register of the emerging multimedia regime, has no specific home state of being. She does not even have a home medium, as her identity as a recording artist is inextricable from her performance in music videos

(Madonna would be nothing without MTV). The "digital lingua franca: bits" (which in her case "sell so well") allows her to change not only her appearance, but also her format, "for as many purposes as possible."[56] Although it may seem at first that Negroponte mistrusts the ways in which this mobility and transformation of data serve the society of the spectacle (a suspicion that will be addressed later in this chapter), he is actually concerned with defining multimedia and positioning himself in its development, not in critiquing commodification or consumption.

A year later, music industry magazine *Billboard* labels 1994 "The Year of the Morph," a moniker earned by the introduction of interactive divisions in all major film and record studios and the proliferation of "hybrid audio/visual/gaming products" with "highly malleable" content.[57] What Negroponte calls multimedia, a term that later gives way to "convergent media," is here touted as a newly mainstream direction in the sharing of soft, girlish data. Aptly, this issue of *Billboard* is published when digital morphing is all the rage. *Time* magazine has just used morphing software to generate a hybrid product with malleable content for its cover: "Eve," the "New Face of America," a now infamous image that digitally combines faces of different races "to create a composite subject of a networked economy" as a symbol of an America with an allegedly increasing acceptance of diversity.[58]

With both invocations of the morph, the digital condition is marked by a perceived openness to outside influence — the music industry invites transmedia commodification in the service of innovation, while white America, the default position so problematically assumed by the picture on the cover, supposedly embraces miscegenation as evidence of its moral and sociocultural development (an openness contradicted by continuing systemic racism). If Madonna and Eve are indeed the best examples of the multimedia morph and multicultural morph, respectively, then they remind us that the plastic image is a girl who is nothing or everything, racially unmarked (Madonna) or racially composite (Eve) — either way, without deeply rooted cultural or ethnic affiliations.

iii.

Being so many different sizes in a day is very confusing.
—Alice, *Alice's Adventures in Wonderland*[59]

[Director Tim] Burton moves his teen Alice through the film like a JPEG within a design program, submitting her to various manipulations and mobilizations. What we get on-screen is a young woman successfully coming to grips with the use of her own self-image, learning from the program how to finally (endlessly) put herself to work. —John Kelsey, "John Kelsey on Tim Burton's *Alice in Wonderland*"[60]

Although Madonna uses the digital lingua franca to maintain her multimedia stardom and Eve owes her life to software, plastic images are not exclusive to digital mediation. I've already discussed the mutability of persona in Cindy Sherman's analog photographs. We can go back much further. Aspects of nineteenth- and early twentieth-century girlhoods anticipate twenty-first-century media formations (as contemporary notions of youth's dynamic lability were shaped, in part, by earlier understandings of the animation and pliancy of the young). In other words, girlishness had a digital nature *even before the digital*, and certain practices of mechanical reproduction give us digital girls avant la lettre. If "New Media does not radically break with the past; rather, it distributes weight differently between the categories that hold culture together, foregrounding what was in the background and vice versa," as Manovich claims, then girlish plasticity should appear in the background of prior regimes of mediation.[61] How and where can we find it?

Alice in Wonderland can be used as a lens—to be more specific, a telescope—through which to view figurations of protodigital plasticity. Indeed, she is one of the most valuable paradigms of girlishness. From her earliest manifestation as one of Lewis Carroll's photographic models to her recent appearance in Tim Burton's *Alice in Wonderland* (2010), she embodies the elastic, metamorphic, and transmedia qualities of the plastic image. Analog Alices do not usually perform these tricks with the greatest of ease. However, it is precisely through the sometimes painful negotiation between the heaviness of flesh and lightness of images that the girl has emerged as a mediator. The following roughly chronological excursion through Alice's history will provide a partial genealogy of convergent forms of the image and self. The importance of such protodigital plasticity does not unravel the interrelationship of girlishness and digital media. On the contrary, it acknowledges the desire for metamorphic images that later media technologies would claim to satisfy better, faster, and more smoothly. Certain analog practices of image production and the theories that seek to explain them cleared a space for software-driven manipulations of the image to grow and then overgrow their bounds.

The most pressing reason to go down the rabbit hole is to bring (analog) cel animation into our discussion of computer graphics. Manovich famously argues that "digital cinema is a particular case of animation."[62] Hence, it is reasonable to look for girlish plasticity in such analog alterations of the image. What does Catmull's wire-frame hand, for example, have to do with the Caterpillar's shape-shifting cartoon smoke rings in Disney's *Alice in Wonderland*? How does Alice's "opening out like the largest telescope that ever was" relate to the Walt Disney Studio's animation process?[63] What connections can be made between the literary Alice and the software application of the same name, designed to teach object-oriented programming? Alice is both a proper name with a full and complex history and an empty line drawing with flexible contours, able to be filled with everything or nothing. She is a theory of girlishness as a metaform of selfhood.

Let me begin with Alice's real life, or, in gaming and social media terminology, her RL. The protagonist of *Alice's Adventures in Wonderland*, first titled *Alice's Adventures Under Ground*, was named after Alice Pleasance Liddell, who, along with her sisters and parents, was a friend of Charles Lutwidge Dodgson (who published under the pseudonym Lewis Carroll). Originally an oral tale to entertain the children during their languid afternoons together, it was written down in 1863 and published in 1865. Carroll's amateurish but charming drawings of Liddell were replaced by John Tenniel's illustrations, which likely took another one of Carroll's photographic subjects, Mary Hilton Badcock, as their model for Alice. Tenniel also illustrated the sequel to *Alice's Adventures in Wonderland*, *Through the Looking-Glass* (1871). Since then, Carroll's works have spun out numerous iterations, translations, transformations, and appropriations — more than other children's books, both because of their appeal and because they are in the public domain. As Helen Pilinovsky reminds us, Alice "has passed from being a specific character to being a near archetype: images of Alice appear in every medium."[64] Lori Waxman maintains that Alice's extraordinary aesthetic and cultural "pliability" is due to Carroll's narrative, driven as it is by transformations of her body, as well as her unstable identity as an adolescent "on the cusp between childhood and adulthood, a liminal zone that leaves

[her] susceptible to being further pulled this way and that."[65] In other words, *Alice's Adventures in Wonderland* is about girlish variability and thus lends itself to recontextualization and transmediality.

Alice, her sisters Edith and Lorina, and a number of other children were frequent participants in Dodgson's experiments in photography, perhaps the author's favorite medium. One of his most memorable images is *The Beggar Maid* (1858), a photograph of Alice dressed in tatters, her hand cupped at her waist (asking for money, albeit without much conviction), and her eyes looking at the camera with mild distrust (plate 2). She is outsized by her dress; it hangs low on one side, revealing part of what would be her breast if she were postpubescent. This and the careful arrangement of a rip in the fabric to expose her shoulder suggest a certain eroticism of his child model, exacerbated by the accusations of later generations that he sexually abused or exploited her and perhaps others. These allegations remain unproven, and thus I won't discuss them in detail. What I can confirm is that Dodgson admittedly preferred to photograph little girls and presented the perceived lability of their personas and the instability of their identities as attractive characteristics through photographs of his models wearing different outfits (and sometimes none at all, in keeping with the Victorian cult of the supposedly innocent nude).[66] There's Alice and Lorina in Orientalist attire, Lorina holding a black doll, several girls lounging on chaises in poses suggestive of prostitutes in nineteenth-century paintings, Alice in a flowered wreath, his little friend Xie Kitchen in Greek and Danish outfits and later dressed as a Chinese merchant and as Penelope Boothby, girls photographed as a male character from a popular play, the Hendersons dressed as Zulus, yet another pair of Orientalist pajamas, and so on.[67] Stephen Monteiro connects "the divergent, even contradictory, portraits of Alice Liddell," in particular, to the dizzying changes in size experienced by the literary Alice.[68]

Carroll's investment in the littleness of his child models, accentuated by baggy clothing or large furniture, impelled him to photograph them again and again. The microsoft girl of the twenty-first century is prefigured by these serial images, their multiplicity yoked to the girls' malleability and relatively small stature. While the fictional Alice is hardly ever the right size, Carroll expresses concern that his subjects might outgrow the frame, as he imagined it, and the RL Alice, transformed into one of the most beloved heroines of English literature, is shrunk repeatedly to the size of a photograph. For Carroll, the dynamic malleability of the girl is best expressed as a series of images of a series of girls in a series of costumes.

The affiliation between Alice's startling bodily transformations and Carroll's many photographs of girls is made clear on the final page of Liddell's own copy of *Alice's Adventures Under Ground*, which has a small headshot of Liddell glued on top of a rejected drawing of the same image (plate 3). Carroll's replacement of the illustration with a photograph is also a transformation of the book into a photo album and the RL Alice into a transmedia figure available for posterity. Monteiro claims that Carroll's two Alice books are touched by "photography's shadow" in other ways as well, "from the opening entry hall with its small, cameralike doorways to . . . the fluctuations in Alice's 'developing' body and personality as well as the flat, playing-card-sized royal court, not to mention the remarkable figure of the Cheshire Cat, whose emerging and lingering apparition recalls the vicissitudes of prints in darkroom baths."[69]

Almost a hundred years after Liddell receives her lovingly handwritten manuscript of *Alice's Adventures Under Ground*, Kodak makes a commercial demonstrating the photo album's ability to capture the girl's unstable identity.[70] While Carroll highlights girlish plasticity through the alteration of persona, this commercial figures transformation as a repetitive process of maturation that undermines the teleological trajectory of childhood development. It begins with an image of an album entitled "Judy." Then it moves to a picture of a nude baby against an almost impossibly large door clearly chosen to emphasize her tininess in the manner of Carroll's furniture. We see little Judy's charmed and charming life in pictures. She grows up before our eyes, becoming a girl in a pinafore, a child who appears strangely Victorian (more Alice Liddell than a twentieth-century American girl), a kid in jeans, an adolescent in a prom dress, and finally a young mother holding a baby. Important for my purposes is a sequence in which Judy is again pictured in front of the giant door and a series of dissolves allows us to see her move through all of the stages of her life in rapid succession. The most interesting points in the sequence are the ambiguous, elusive, and blurry in-between moments when two images of Judy cross-dissolve. What this commercial does is take the photographic series to the point of cinematic special effect (at this point, a not-so-special effect), fulfilling the promise of the photo album through other means. Kodak shows us that a book of photographs can both pluck the special moments out of the flow of a girl's life and, through what is not a book of photographs but a cross-dissolve, give us the pleasure of watching and rewatching the (not quite) smooth flow of developmental transformation. This commercial can be compared to a recent advertisement for Sprint that gives us four photo albums for the Information Age—that is, four smartphones.[71] Once again, we see a girl develop in a series of pictures. However, this time she doesn't become a mother. This girl stops growing at early adolescence. The commercial recaps her transformation just in case we missed it the first time—as did the old Kodak commercial—and still, she ends up a tween. Sprint's tagline is "Unlimited phone. Unlimited data. Unlimited life," folding the possibilities of data together with the potentiality of girlhood, the animateness of the image with the girlish vivacity of information. Here the girl encounters not a looming door, as Judy and Alice do, but a giant hand, huge in relation to the grinning face on the small screen of the smartphone. The life-giving hand of the animator, a trope of early twentieth-century animation, reemerges in this commercial as the nimble parental fingers gripping the screen within the screen.

Alice's history on screen is almost as old as the history of cinema, itself. Three years before the first animated film—and the first hand of the animator appeared within the frame—we have the inaugural film adaptation of *Alice in Wonderland* (1903).[72] The lively metamorphoses of Wonderland proved inspirational and challenging to Cecil Hepworth, Percy Stow, and other early filmmakers who wanted to use special effects, harnessing the attraction of girlish plasticity to technical innovation.[73] Her most compelling and indeed strangest early appearance is in Walt Disney's *Alice's Wonderland* (1923), the pilot film for a series of breezy short subjects made in the 1920s. With only the most tenuous connection to *Alice's Adventures in Wonderland*, the *Alice* comedies feature a child actress interacting with various cartoon characters in both live-action and animated environments. Carroll's novel and Disney's Alice shorts are connected not through narrative but by the basic premise of Wonderland as a stage for girlish transformation. The animatic process involved pasting stills of the "real" actress onto animation cells; Alice's lateral trajectory from live-action frame to animation cel replaces the vertical movement of her descent into Carroll's Wonderland.

Alice is played by four different girls throughout the series' four-year, fifty-eight-film run — four-year-old Virginia Davis, six-year-old Dawn O'Day, four-year-old Margie Gay, and the older Lois Hardwick. Historiographically, Gay's presence in the series unfortunately overshadows that of Hardwick, the final and, according to Disney, the best Alice. As Russell Merritt and J. B. Kaufman explain in *Walt in Wonderland*:

> And one identification note. Amidst our research, we were startled to learn of the existence of an uncredited Alice who over the years had been turned into a non-person. The last ten films attributed to Margie Gay in fact starred someone else, a young girl who was several years older than Margie and a far more expressive actress. . . . We have been able to identify this charming youngster as Lois Hardwick.[74]

Hardwick is often mistaken for Gay in a photograph of the former posing in contrapposto among Disney and his staff.[75] Though this may be the effect of my knowledge of her historiographic erasure, I can't help but notice that Hardwick appears less self-assured than her counterpart does in a comparable image, as if she has a vision of her future as a nonperson.

The Alice of these short films is less a character than a symbol of the continual mutation of the girl-body: spliced into different cartoon worlds, her image subjected to various cinematic and animatic effects, the character transforms into four very different girls as the series progresses. She is eminently replaceable, transmedial, and transportable (from literature to film, from live action to animation, from "real" worlds to dream worlds). And in the case of Davis, at least, she is also posable. A behind-the-scenes image from *Alice the Peacemaker* (1924) shows two assistants arranging Davis's body to be photographed. She is pictured holding a running pose: one assistant positions her left leg back and bent at the knee, the other adjusts her hip.

The inviting reconfigurability of the plastic image (and the girl who is made to embody it) has influenced not only popular culture, but also the historical avant-garde. Most applicable to our collection of Alices is the work of German-Polish surrealist Hans Bellmer, who constructed and photographed two near life-sized adolescent dolls in various poses and environments. Bellmer's series of doll photographs from the 1930s present the girl as a mutational, variable plastic image and anticipate postphotographic malleability.

The first set of these images is published in a limited-edition book entitled *Die Puppe* (*The Doll*, 1934). According to Therese Lichtenstein's evocative description:

> As the viewer turns each page, moving from one image to the next, an atmosphere of apprehension arises; the restrained yet flexible doll, bathed in dramatic light and photographed from different angles, is transformed in successive shifting images to create a sense of temporality and ephemerality, not unlike that produced by a series of film stills.[76]

These photographs portray an unstable body in various morphogenic states revealed through successive images. A characteristic photograph of the first doll, a four-and-a-half-foot-tall plaster and papier-mâché construction over an armature of wood and metal, depicts her leaning against a wall and glancing coyly over her shoulder. She wears a wig and is missing an arm; one of her legs has no plaster flesh, only the bare projection of the wooden "bone." Another photograph shows her completely disarticulated, with her component parts—torso, head, legs, wig, ball joints, and a single eyeball—arranged like a ritual offering on the striped ticking of a mattress.

The second doll has a more bulbous form; Bellmer created a central ball joint to represent the abdomen, around which the hips and upper torso could easily rotate. This doll was often photographed with multiple limbs, hips, buttocks, or breasts. In one representative image, she has four legs and two sets of hips and genitals. Leaning against a tree, she stands in a casual contrapposto position, one leg bent at the knee, one hip slightly lower than the other (plate 4). To the right of the doll is a mysterious, partially concealed hole in the forest floor. The void in the earth looks like a rabbit hole.

The dolls and the photographs that feature them are plastic forms enacting "the self in its constant state of mutation, multiplication, and division."[77] Producing a notion of girlhood as ontologically bound to corporeal reconfiguration, they contribute to the burgeoning visual discourse of plasticity. Bellmer later explains that "the body is comparable to a sentence that invites you to disarticulate it, for the purpose of recombining its actual contents through a series of endless anagrams."[78] Of course, he is not rewriting *the* body, but a specifically girlish one made available as an expression of the generic body-in-process.

These are painful, cruel, and deeply unsettling images. The girl is the locus of a disarticulation, an unraveling of language, which is also a seemingly never-ending process of dismembering and remembering (an un/doing that reaches its fuller potential with digital media). On the other hand, a convincing reading of these images is that they are not themselves sadistic, but reflexive "representations of sadism"— that is, the "sadism of fascist father and state."[79] Bellmer's work troubles the steeled, inviolable monolith of the fascist male body by replacing that image with photographs of two variable dolls captured in the fleeting moment of their mutation. Fascist Germany eradicated those citizens who did not fit the image of a compliant heterosexual Aryan populace with "healthy" bodies; Bellmer responds (or rather presponds, since the full plan of the Nazi regime was not yet entirely clear when Bellmer began photographing the dolls) with degenerate girls.

Around the same time that Bellmer publishes *Jeux de la poupée* (*Games of the Doll*, 1949), a book of some of his doll photographs from the 1930s that are later hand-colored by his friends, and the Disney Studio is inking and painting images of another bow-headed adolescent, artist Henry Darger is making plastic images to illustrate his own book: a 15,145-page epic with drawings and collages of little girls. *The Story of the Vivian Girls, in what is known as the Realms of the Unreal, of the Glandeco-Angelinnian War Storm, Caused by the Child Slave Rebellion* occupies him for decades. Influenced by histories of the American Civil War and children's fantasy literature, *In the Realms of the Unreal* (as it is conventionally abbreviated) tells the story of enslaved white children and the heroic Vivian Girls, leaders of a successful child rebellion. The illustrations are recognized for their remarkable use of watercolor, dramatic horizontal orientation, ornate compositions, and depictions of literally gut-wrenching acts of violence against children — all produced by an artist lacking traditional formal training. Without explanation (although the subject of fierce scholarly conjecture), Darger gives some of the girls small penises (plate 5).

Darger's work is an important addition to our genealogy of girlish plasticity. A sophisticated and prescient consumer, he appropriated pictures from comics, coloring books, newspapers, and advertisements, and either traced them or, less frequently, collaged them into his work. He modified the original images by adding horns, tails, and props, changing clothing or removing it, and penciling in male genitalia. His mature process elegantly demonstrates protodigital scalability: Darger enlisted the help of his local drugstore photo studio to enlarge or shrink the found images he later traced (the resulting changeability and disorienting scale of the figures and environments evoke Carroll's Wonderland). In consideration of Darger's active mode of consumption and fixation on images of fighting girls, Tamaki Saitō labels him as an American version of a Japanese *otaku* (a passionate fan of anime and manga, usually assumed to be male).[80]

Many of Darger's figures have similar faces and bodies as a result of the visual consistency among images of girls in early to mid-century American popular culture, which was still working through visual motifs of childhood innocence established in the nineteenth century. More importantly, this repetition-with-a-difference is also due to Darger's practice of keeping a realspace database filled with pictures of girls with different expressions, poses, and hairstyles and then using the same pictures over and over again. Instead of individualized characters, *In the Realms of the Unreal* gives us a relatively small number of ur-figures multiplied and modified to generate a legion of girls. Thus Darger's process, although radically different in intention and much less precise, is similar to computational practices that "produce an entire class of drawings . . . running through a specific pattern in all its variations."[81] The mutability of Darger's figures is even more striking when we consider the otherwise mysterious transgendering of the Vivian Girls. As Frenchy Lunning writes of the Japanese *shōjo* (an adolescent girl), but equally applicable to Darger's cuties, as "the most vulnerable and undervalued" subject she is "easily transformed into other genders and beings that appear inconsequential to mainstream cultural meanings and agendas."[82] Little girls can be anything, Darger seems to say, even boys (and, for that matter, fantastic girl-faced, winged creatures he calls "Blengins").

The plasticity of Darger's work facilitated the reuse of some of his images for political satire in the aftermath of a 2011 student demonstration at the University of California, Davis. Through a technique not unlike the one used for Disney's *Alice* shorts, a photograph of a police officer deploying pepper spray on a group of peaceful student protestors was digitally collaged into various famous paintings, illustrations, photographs, film stills, and screen captures of video games. While the number of possible source images is almost endless, I am struck by the many Darger illustrations that have been reconfigured for this meme, especially considering the fact that his works are not as well known as the majority of the other Photoshopped images (most of which insert the officer into canonical paintings or stills from popular cartoons). Not only the drawings' thematic concerns regarding the oppression of the young, but also their status as already appropriated and recontextualized images, invited the anonymous participants in this meme to repurpose them to critique the abuse of power. Consequently, Darger's work, already soft, becomes even more malleable; state power also becomes something less rigid, something manipulable, through the scalability of the officer's form as it is made to fit into different compositions. Not surprisingly, many other collages are based on Disney cartoons, casting the officer in the role of the dastardly villain (and taking a swift jab at the Disney corporation, with its notoriously rabid response to copyright violation).

While it is unclear whether Darger was directly influenced by the look of Disney heroines, his source images and the early animations of the Disney Studio both drew from the same popular iconography of childhood — beribboned, big-eyed, often chubby-cheeked girls visually associated with oversized blossoms, butterflies, and cutesy animals (although Darger combined this anodyne imagery with muskets, eviscerated torsos, and scenes of war). Disney produced another version of this archetype for the feature *Alice in Wonderland* (1951), starring child actor Kathryn Beaumont as the voice and live-action double of the animated character. Like adolescent dancer Marge Belcher before her, who modeled the gestures and choreography of Snow White and some of the dwarfs, Beaumont and all of the other voice actors were filmed to ensure the believability of the animation (as Darger chose to trace others' illustrations). The Disney Studio followed a rotoscopic process developed for *Snow White and the Seven Dwarfs* (1937): live-action movement was recorded, each frame printed as a photostat, and the gestures traced and exaggerated. While Disney animators Frank Thomas and Ollie Johnston are adamant that her choreography was not copied exactly, Beaumont claims nonetheless to "recognize some of the movements" in both *Alice in Wonderland* and *Peter Pan*, for which she modeled Wendy. "Back to Disney Studio," an article written about Beaumont's return visit to the studio in 1983, claims that though she is no longer the "vivacious" child she once was, "the presence of Wendy and Alice can still be detected in an occasional inflection or a gesture of her fine-boned hands."[83]

Whatever naturalism was achieved in the animation of *Alice in Wonder-*

land was secured through very unnatural means.[84] "Back to Disney Studio" describes how the young Beaumont "had to clamber into a large plastic contrivance that represented the bottle floating on a sea of tears, and ride in it while it was rocked on a dolly."[85] A production photograph depicts another mediating construction: Beaumont is propped up on a wooden stand, her feet dangling and her arms held up and bent at the elbow, her white pinafore lifted by two ropes. While the animated Alice falls gently, almost weightlessly down the rabbit hole, Beaumont's weight must be held up by a stand. "The perfect prototype of Alice," as a *Los Angeles Times* article describes Beaumont, nonetheless falls short of "her cartoon twin," and Beaumont's live-action game of pretend is, in a sense, more contrived than the resulting animated film.[86]

Disney's series of *Alice* shorts and the production photographs for *Alice in Wonderland* stage the figure of the girl as both bound to and moved by her animated image. The ropes, wooden stands, plastic constructions, (male) assistants and animators — all devices of mediation — forge the link between girl and image; at the same time, they expose the discomfort of that continuing discursive and material bond. Without a doubt, these contrivances aggressively constrain the girl; however, they also facilitate her transformation into the flexible, mobile plastic image that she supposedly already is. In other words, Alice the animation realizes the innate potential of Beaumont the prototype.

We can interpret the material difficulties of Beaumont's attempts at weightlessness as proof that girls are not images. Not *really*. Indeed, they are not. Bellmer's dolls and Darger's source materials lent themselves to reconfiguration through photography because they were always already images, while Beaumont was a girl IRL. However, I want to argue that despite the physical impossibility of fully embodying Beaumont's cartoon self, aspects of the actress were still taken, reformatted, and shared, and the attachment between the real and virtual girl was still documented and disseminated through mass media. Moreover, these paratexts supporting *Alice in Wonderland* highlight Beaumont as the film's primary representative of our desire to become a plastic image, despite the fact that the studio regularly rotoscoped men and adult women, not just girls as chronologically defined. Alice is the heroine of the film, of course, and a focus on her is understandable. Even more important than this, though, are Alice's and Beaumont's roles as figures of identification for viewers who themselves wish they could enter the fantastic and terrifying world of Wonderland by becoming a plastic image.

If Beaumont is Alice's prototype, then she is also the prototype of the digital avatar, "a prosthetic icon available to anyone, regardless of their [*sic*] nominal identity."[87] Avatars allow users to explore virtual spaces. Likewise, Alice the literary heroine mediates between the real world aboveground and the fantastic realm below. She performs a similar role in Tim Burton's *Alice in Wonderland* (2010). Actress Mia Wasikowska, as the character Alice Kingsleigh, is able to travel between a photorealistic environment and the garish digital fantasy underneath. As she moves throughout Wonderland, she also stitches stereoscopic 3-D to conventional cinematic images: the unsophisticated 3-D effects transform the picture into a series of flat planes that Alice, always the intermediary, can move between. Alice is what allows the film to hold together visually. She is an interdimensional traveler whose body can adapt to different mediascapes and whose wanderings trace out a kind of multiplatform identity.

These skills are put to good use by the makers of the software application Alice, which introduces students of computer science to object-oriented programming. The face of the educational product is a doe-eyed cartoon girl clearly designed to conjure up Carroll's heroine (with a touch of anime). She has grown so big that she now holds the whole globe in her palm, and was explicitly created as an accessible, inviting vehicle allowing students to enter the world of computer science by programming 3-D wonderlands. The creators of the program's "entertainment-oriented" pedagogy assert that it has the potential to change the face of the field of computer science.[88]

Alice is always brokering something.

The most explicitly political use of Alice's transmedial, transformational figure — and her concomitant ability to sew together worlds of different dimensions (classrooms and virtual computational spaces, in the above example, and 2-D and stereoscopic 3-D landscapes in the previous one) — is La Tana di Alice, part of proposed extension to Pinocchio Park in Tuscany, called "The Park of Pinocchio's European Friends." Designed to further the aims of the European Union, a mission described on the park's website as promoting "the representation and integration of European cultures," the

Alice Pavilion will join other themed architectural structures based on the adventures of various European children's book characters.[89] The entire park is conceived as a "hypertext." As the website explains (somewhat elliptically), "By moving from a main story-line (the walk through the Park) through various locations (events), totally different worlds and stories can be discovered."[90] The world particular to La Tana di Alice is based on digital image manipulation and projection, and the principal discovery is the supposed fluidity of identity and affect in the new Europe. Principal architect Lars Spuybroek explains:

> To get inside the *Tana di Alice* we have to go through a hole (like Alice), and after passing through an underground tunnel we enter the complex formed interior . . . In the tunnel there are cameras where visitors can have their faces downloaded into the Replicator Program, which changes their emotion, sex and age. The new faces will be projected in real time in the Mirror Hall at several different zones in the space and the interior's highly reflective surfaces will provide an infinite number of reproductions.[91]

The Replicator Program "morphs men into women, alters the emotional expressions on their faces or changes their age."[92] Here Alice's falling, shrinking, and enlarging—as well as the mobility of her identity—are staged through the computergraphic technology of the morph, the mainstay of digital animation and cinematic special effects.

Visitors' morphing faces reinforce the transformation of the geopolitical face of Europe, the metamorphic play of Self and Other that has characterized the fraught consolidation of the European Union. Bringing together different scales of transformation, Spuybroek's designs for La Tana di Alice also support Vivian Sobchack's claim that "the morph, in the microscopic fluidity of its figural transformations, enacts the more macroscopic and fluid quick-changes not only of the computer and entertainment industries that generated it but also of the larger metamorphic technosphere in which we all live."[93] Sobchack's groundbreaking edited volume on the morph, aptly titled *Meta-morphing: Visual Transformation and the Culture of Quick-Change*, presents morphing as both a kind of digital effect and a "*cultural imaginary* that links a present digital practice to a much broader history and tropology of metamorphosis and its meanings."[94] The proposal's interactive Mirror

Hall jumps scale with each infinite reflection; the fluid features of morphed faces perform the assumed fluidity of national borders within the EU. The geometry of the pavilion, as Spuybroek envisions it, moves "between the idea that change emerges from the self and that it is enforced from the outside"; images of personal, cultural, geopolitical, and racial change are projected on its rippled interior surfaces.[95]

What does it mean "to go through a hole (like Alice)"? What is reflected in the long Mirror Hall of Alice's history?

First, as we have already discovered, she is transdimensional, transmedial, and cross-platform. Her skills lie in her ability to move between spaces or worlds (such as the nation and the Union, 2-D and 3-D filmmaking, live-action and animation) and to embody technological innovation and social change. Second, she is a subject-in-the-making. Her mutability and scalability reinforce the processual nature of protodigital images and identities. If girlishness can be done, it can also be undone and done differently (to use computational terminology, girlishness has unlimited levels of undo). Any individual girl in Darger's drawings can meet a gruesome demise and then reappear in another panel, larger or smaller and with a new hairstyle, a penis, or wings; she can later be reborn as the victim of pepper spray.

Concomitantly, Alice is a serial creature. Any single manifestation would wither if left alone; her animation requires repetitive acts of mediation. In other words, Alice Liddell and the other girls' transformations are only truly legible if they follow the logic of the series. While Carroll's photographic experiments obviously had to end eventually, when their models aged out of childhood or their photographer aged out of life, they still promise to perform the girl's vivacious onwardness into a seemingly infinite future. If she has no home state of being, her identity can follow a vector forever and in any direction, and her Self can be reconfigured into any number of Others. Likewise, La Tana di Alice — if built — would attempt to produce the present of the EU by moving into a future in which any additional visitor to the pavilion can morph into anybody else ad infinitum (otherwise, the limitations of the EU would be exposed). The plastic image comes into being as a series of numbers — let's say, prime numbers. There is always another prime, and the girl is always in it.

iv.

Immediately the animators tried to outdo each other in making drawings with more and more squash and stretch, pushing those principles to the very limits of solid draftsmanship. — Frank Thomas and Ollie Johnston, *The Illusion of Life*[96]

In the Cartoon, which is one of the most popular and in some respects the best medium of cinema expression, the human atom and its belongings, undergo whimsical changes that cause a continuous stream of images to form in the mind . . . But the Cartoon never departs from the actual. It consists of an elastic line in evolution. Shapes grow out of it with which we are familiar even though they are distorted and battered by a sort of recurrent earthquake. In other words, the Cartoon of the Mickey Mouse, the Krazy Kat, the Felix the Cat . . . is simply the caricaturist playing with line that has the elasticity of gas. It shrinks and expands, collapses and recovers, behaves like a spring winding and unwinding, and at the same time assumes the shapes and characteristics of human beings, animals, insects, of animate things, and inanimate ones made animate. These extraordinary puppets of all sorts, that fall to pieces in heaps and reunite, and outdo even an India-rubber ball in diversity of shapes . . . have a distinct sociological value.
— Huntly Carter, *The New Spirit in the Cinema*[97]

Digital film has split the atom.
— David Clark, "The Discrete Charm of the Digital Image"[98]

Animation theory, from Huntly Carter's *The New Spirit in the Cinema* (1930) to Scott Bukatman's compelling book *The Poetics of Slumberland* (2012), has long recognized the malleability of the image — what I call its girlishness, its Aliceness. A characteristic description of the ontology of animation comes from Paul Wells: "The mutability of the body, and indeed, the whole material environment, is a fundamental aspect of the way animation revises and questions aesthetic norms and social orthodoxies."[99] For the most part, the qualities of lability and elasticity in animation are understood as positive and often resistant, marking the field and its understanding of protodigital imagery with a certain girlphilia. Take Sergei Eisenstein's ruminations on Disney cartoons. An exploration of the utopian potential of girlish plasticity, this work is arguably the most formative contribution to animation studies. And whom do we see but Alice yet again? — Eisenstein offers one of Tenniel's illustrations (see figure 1.3) as evidence of the long-standing appeal of dynamic forms, which he famously describes as "plasmatic." He comes to the conclusion that it is the plasmaticness of the animated figures in Disney cartoons, their "ability to dynamically assume any form," that establishes their nature, gives them their power, and elicits what he considers a primal desire to see and produce malleable images.[100]

Eisenstein's oft-cited conceptualization of the plasmatic, "a rejection of once-and-forever allotted form, freedom from ossification" (that is, "the transformation of stable forms into forms of mobility"), has been taken as essential to the ontology of animation, regardless of the particular technique.[101] However, plasmaticness is not limited to animation proper. As I follow Alice through time and space, Eisenstein chases mobile forms across media, finding them in Disney's films and children's book illustrations; in the movement of water, clouds, and fire (the last of which is compared through a literary reference to dancing women); in the circus act known as the "human snake"; and, with his most primitivist comparison, in the movements of "the 'Snake Dancer' of New York Negro nightclubs, where the same kind of creature writhes in abstract, silk robes."[102] His text maps out an expanded notion of animation that is both motivated by early twentieth-century primitivism and remarkably applicable to our own hyperanimated contemporary media cultures. Influenced by Eisenstein, Bukatman also tracks plasmaticness across media: newspaper comics, comic books, the films of Jerry Lewis, the children's book *Harold and the Purple Crayon* (1955), the documentary *Le Mystère Picasso* (1956), and so on.

Eisenstein's writings establish a strong associative relationship between plasmaticness and feminized, infantilized, raced, and nonhuman bodies. Indeed, the plasmatic is intimately entwined with issues of agency as they relate to girl-subjects and other marginalized beings. For example, Eisenstein attaches an affect to the consumption of plasmatic forms — ecstasy, or "a sensing and experiencing of . . . the 'plasmaticness' of existence, from which *everything* can arise" — and then uses a psychiatric case study of a girl arsonist as the primary model of ecstasy's scopic dynamics.[103] Following repeated comparisons between plasmatic forms and the "eternally changeable" dance of flames (which magnetizes the viewer with "the variability of its living forms"), he claims that obsession with fire is especially common among "young girls in the period of *Pubertät*."[104] "In their blood burns the *fire* of desires," he quotes from an uncited source, arguing for the consonance between the girl's physiological/affective state and the images-in-motion that she both contains and observes.[105] Although Eisenstein's concepts of plasmaticness and ecstasy are not the girlphobic theories of Siegfried Kracauer and Marshall McLuhan, discussed later in this chapter, they do share an insistence on the immaturity of the viewing subject in the face of the spectacle. However, while McLuhan and Kracauer show great concern for the girlification of the image consumer, Eisenstein sees the radical potential, however temporary, of both the underdeveloped subject and the plastic image.

Eisenstein focuses on a case study of a fourteen-year-old female arsonist as evidence of the sensuous and affective allure of fire's ever-changing forms. Quoting from a 1941 book titled *Legal Psychiatry*, he explains that the little pyromaniac's "desire to commit arson arose in periods of a sort of unaccountable melancholy and inner unrest, during which she became irritable. It was then that the desire to see fire would arise." Suffering from irrepressible urges, she "first noticed her love for fire" at the age of eight; by fourteen, she exhibited "physical underdevelopment" and "an organic affection of the central nervous system" resulting in pyromania.[106] In other words, she expresses her inner physical and mental turbulence by choreographing a dance of restless images. After all, the etymology of the word "ecstasy" combines *ek* (Greek for "out") and *stasis* (or "place"); she is out of place, displaced, outside of herself. According to Eisenstein, the fire of her pubescence (or protopubescence) becomes externalized as a spectacle of protean shapes.[107]

Although Eisenstein chooses an adolescent criminal as the poster child of the ecstatic viewing experience, he is sympathetic to her love of fire's plasticity. Eisenstein wants us all to be impulsive girl pyromaniacs. His characteristic editing style, the montage of attractions, reflects his desire to elicit spontaneous psycho-physiological reactions in the viewer. Eisenstein hoped that the visceral, vital, primitive impulse would move spectators of his films to change ossified socioeconomic institutions. Early twentieth-century Americans, especially, needed (and loved) Disney's plasmatic figures because of the oppressive inflexibility of modernity in the United States — Americans' "mercilessly standardized and mechanically measured existence."[108] Plasmaticness temporarily returned to the viewer what the assembly line had stolen: "a lost changeability, fluidity, suddenness of formations."[109] In other words, the teenaged arsonist, who was underdeveloped even for her age, retained what Fordist societies sought to stamp out through the development of their citizens. Here Eisenstein disagrees with an interpretation of the Fordist/Taylorist subject later advanced by Latham. The former believes that plasticity challenges the rigor mortis of factory life, the latter that it gives capitalism what it needs to survive.

Pathologized, underdeveloped, and criminal — that is, unfit for labor — the adolescent arsonist is a non-normative subject with an inability (or, I would say, lack of desire) to maintain the boundaries between self and image; her transgressive identity confounds the "timelock mechanism of American life."[110] She joins other marginalized subjects and feminized nonpersons featured in Eisenstein's writing — adult women, African American dancers, circus performers, children of both genders, fetuses, animals, waves of water, clouds, and, of course, fire — whose metamorphic power offers the mature, white, masculine spectator-worker a fleeting vision of minoritarian freedom. These malleable, rubbery, mobile, and energetic bodies do not go by the name Alice. However, Eisenstein would say that they all have ingested the same liquid that makes Alice shrink and expand.

Carroll's Alice isn't a criminal on par with Eisenstein's arsonist. She is unjustly accused of eating the queen's tarts, after all. Importantly, though, the novel presents her as unsuited for work. When she begrudgingly follows an order to fetch something for the White Rabbit, who mistakes her for his housemaid, she complies, but then grows so big that she can't exit through his door. Alice's metamorphic body prevents her from being a productive laborer or, for that matter, wife. While Eisenstein's pyromaniac burns down houses, Alice prevents their owners from occupying them.

Alice and the arsonist are both sisters of Paul Virilio's child picnoleptic, who is also defined by her love of plastic forms. Virilio's concept — which is a theory of animation without being recognized as such — is built on his reading of the stop trick. The basis of metamorphic effects in stop-motion animation, the stop trick was allegedly discovered by George Méliès, considered to be the progenitor of a long line of films that foreground fantasy and magic, including most animation. Méliès describes his Eureka moment in the following passage:

> One day, when I was prosaically filming the Place de L'Opéra, an obstruction of the apparatus that I was using produced an unexpected effect. I had to stop to free the film and started up the machine again. During this time passersby, omnibuses, cars, had all changed places, of course. When I later projected the reattached film, I suddenly saw the Madeleine-Bastille Bus changed into a hearse, and men changed into women. The trick-by-substitution, soon called the stop trick, had been invented, and two days later I performed the first metamorphosis of men into women.[111]

In Méliès's stop trick, the loss of time created by the faulty cinematic apparatus permits a metamorphosis fitting for a history of animation: men morphing into women. Méliès, invested in his own role in "fathering" the trick film, writes a genesis story that casts himself as both inventor and discoverer of a technique of transforming, transgendering, and turning the "prosaic" into the poetic.

Virilio describes the stop trick specifically and the perception "of malleable forms" more generally in terms of what he calls "picnolepsy."[112] A kind of brief blackout, either accidental or produced through play (think of the song "Ashes, ashes, we all fall down" and its accompanying choreography), picnolepsy needs the closing and opening of the eyes as the stop trick needs the stopping and starting of the camera. The phenomenon is common in children, he writes, but in adulthood, it "stops being mastered and enacted."[113] The adult's loss of picnoleptic competence can lead to an unhealthy obsession with the "technical prostheses of mediatization (radio, motorcycle, photo, hi-fi, etc.)," an addiction to what is only an enervated manifestation of a once dynamic capability.[114] In other words, picnolepsy is like a zombie; it is a jovial friend who returns as a monstrous, heaving creature.

While Méliès "acted like a child" when he created malleable forms, we assume that Virilio means to emphasize the word "like," as he claims that adults can only try (and fail) to regain their childhood "savoir-faire."[115] Here Virilio walks a thin line between infantophobia and infantophilia. There is a disruptive power in the child picnoleptic's playful animation of the world, not unlike Walter Benjamin's understanding of the child's mimetic faculty. The adult who chases this thrill through technical prostheses, however, is immature, even dangerous. In other words, freedom for the child is addiction for the adult. It appears that Virilio does not see eye to eye with Eisenstein, for whom acting like a child has political force and childhood savoir-faire is the viewer's and the filmmaker's most meaningful pursuit.

V.

1. Close any programs that are running.
2. Click **Start**, click **Run**, type regedit in the **Open** box, and then click **OK**.
3. In Registry Editor, expand one of the following registry subkeys, as appropriate for the version of Excel that you are running:

   Microsoft Office Excel 2007

   HKEY_CURRENT_USER\Software\Microsoft\Office\12.0\Excel\Options

   Microsoft Office Excel 2003

   HKEY_CURRENT_USER\Software\Microsoft\Office\11.0\Excel\Options

   Microsoft Excel 2002

   HKEY_CURRENT_USER\Software\Microsoft\Office\10.0\Excel\Options

   Microsoft Excel 2000

   HKEY_CURRENT_USER\Software\Microsoft\Office\9.0\Excel\Options
4. On the **Edit** menu, point to **New**, and then click **DWORD Value**. Select **New Value #1**, type **UndoHistory**, and then press ENTER.
5. On the **Edit** menu, click **Modify**.
6. In the **Edit DWORD Value** dialog box, click **Decimal** under **Base**. Type a value between 0 and 100 in the **Value** box, click **OK**, and then exit Registry Editor.
7. Start Excel. Excel stores an undo history for the number of actions that you specified in step 6.

—Microsoft, "How to Modify the Number of Undo Levels in Excel"[116]

A hairdresser created a poof
So high that it looked like a goof!
He said while a-tugging,
Madame, I am debugging!
This artless and curleycued roof!
—IBM et al., Definition of "Debug"[117]

Eisenstein's girlphilia runs contrary to a discursive trajectory of media studies and visual culture studies, which, as I argue with greater care in chapter 2, are underpinned by an unacknowledged girlphobia. To give a high-profile example, Siegfried Kracauer's "The Mass Ornament"— essential to the field of visual culture studies (among others)— presents the shifting configurations of the Tiller Girls' choreography as both expressive and symbolic of the viewer's commodification. Kracauer's subject is implicitly figured as an adult male threatened by the stultifying force of girlification in the face of capitalism's global spectacle. Marshall McLuhan, often called the father of media studies, writes a whole book about the horrific vacuity of the girlish image, what he calls the mechanical bride. She appears in different forms — girls pushing Coke products, Ziegfeld dancers, smiling drum majorettes, legs in pantyhose, pictures of the globe sucking on a soda bottle — but is always essentially the same: a girl-commodity whose flocked, touchable skin and soft-focus veil belie the danger of her robotic heart. Ultimately, the mechanical bride transforms the adult male viewer into a version of herself: immature, dependent, inhuman, both consumer and consumed. If the figure of the girl embodies an "unfinished process of personal development," then McLuhan's and Kracauer's deeply adultist concerns hold tightly to a notion of the subject as stable, invulnerable, individualized, and complete. Tamaki Saitō observes a similar attitude underwriting harsh judgments of Japanese otaku. "Surrounded by these infantile women"— that is, women who love toys and adopt childish personas — Japanese "men become infantilized as well."[118] Writing twenty years before Saitō, Naoto Horikiri identifies girlification as no less than the Japanese national condition: "We are no longer the shabby and middle-aged teacher Humbert Humbert who chased Lolita's rear-end in his dreams. We all have become the forever-young Lolita herself. We are driven night and day to be relentless consumers . . . [The adolescent girl] has today commodified everything and everyone."[119]

Likewise driving adults and boys to become relentless forever-young consumers, McLuhan's mechanical bride is not only a bad daughter, but also a corrupt "girl-wife" and "darling mother" who turns consumers into infants through the pap that she pushes.[120] A more recent example of media theory that implicitly blames the girlish mommy for improperly nourishing subjects is Vito Campanelli's 2010 book *Web Aesthetics: How Digital Media Affect Culture and Society*. Through a remarkable array of figures of speech, Campanelli calls the Internet a "welcoming womb" and "the breast that feeds us as television had done for decades."[121] At the same time, he writes, the Internet "sterilize[s]" thought and "shape[s]" the denizens of the "network society" with its "castrating forms."[122] It also promotes "carefree," "trendy," and "cool"— dare I say, adolescent — ways of being.[123] In sum, Campanelli's Internet transforms users into an impossible creature: a mutable fetus who is also a nursing infant, trendy teen, and sterilized adult.

Contemporary philosopher of media technology Bernard Stiegler offers a less nightmarish contribution to this conversation. While concerned with age and not gender, his provocative but phobic understanding of immaturity dovetails with long lines of thought that figure women as extending the adolescent phase further than men (as G. Stanley Hall claims in "The Budding Girl"). Note that in English, "boy" is not the exact counterpart of "girl." While the latter is often welcomed by women, even when spoken by strangers (e.g., "The girl over there wants this shirt in a different size"), the former is rarely used to address adult men in similar situations. "Boy" is a more regulated word when applied to men beyond their teens, appropriate when used by lovers or close comrades ("These are my boys" or "home boys") but potentially aggressive when said by others, especially given the racist application of the word. For the most part, boys become men, or at least "young men," sometime in their late teens or early twenties, while women can be called girls for much if not all of their lives. While Stiegler himself does not argue that women are more immature than men, Western cultures have a history of associating women with the same attributes that Stiegler attaches to the immature (and dependent, unthinking, fickle, and impressionable) subject of late modernity.

Stiegler blames commodity culture for eroding the boundaries between childhood and adulthood, infantilizing adults and prematurely shoving the young into mature life. The sophisticated rhetoric of *Taking Care of Youth and the Generations* (his 2010 book on education and media) masks the lurking infantophobia that drives his conception of the knowledgeable, focused, literate, and above all mature subject we all should strive to become.[124] Although his insistence on the importance of institutions of care and critique of Michel Foucault's neglect of consumption are well played, his blinkered rhetoric tends toward the crotchety; I can hear the faint cries of "Kids today!" as I turn the pages. More importantly, his focus on the juvenilization of the adult through the media's shortening of our attention spans — which begins with the advent of radio and then continues with television and is exacerbated by networked media—posits the child as the pathology forced upon the consumer. Indeed, attention deficit hyperactivity disorder stands for our lack of thoughtful commitment to the institutions that support us and will sustain our successors. While I am happy to join with Stiegler in his clarion call for long-term thinking, I am reluctant to discount the power of ephemerality, of the "deficit" and the "disordered," and unwilling to call psychosocial transformation a form of maturation. Stiegler's otherwise compelling understanding of individuation as an ongoing process of change leaves an opening for an enabling reading of immaturity that he fails to give its due.

While Stiegler's media philosophy presents digital technologies as radically accelerating but not initiating the infantilization of the consumer, Alan Kirby's *Digimodernism* implies that the immaturity of digital and especially networked cultures separates our contemporary situation from earlier regimes of mediation. Kirby's aim is to explain the replacement of postmodernism by what he calls digimodernism. His historiography credits or, rather, blames CGI animation for successfully adopting postmodernist aesthetics but ultimately "undermin[ing] postmodernism" by relegating it "to a child's plaything."[125] After kiddie po-mo was discarded as a dominant aesthetic in big-budget animation, especially in the work of Pixar and Dreamworks, postmodernism endures further "humiliat[ion] to become that infant's discarded toy, grown out of and left behind. When postmodernism turned into yesterday's style in the eyes of the children, it surely entered the absolute past tense of contemporary culture."[126] Although Kirby has many fine readings of key works of animation, which is often relegated to the margins of culture, his story of postmodernism's degradation continues the legacy of infantophobia in media studies.

Nonetheless, Kirby's explanation of the digimodernist text, "which displays a certain body of traits that it bequeaths to digimodernism as a whole," resonates with the case studies I discuss in subsequent chapters.[127] His notion of "onwardness" is defined as the digital object's existence as something "coming into being, as something growing and incomplete."[128] "Haphazardness" designates "the permanent possibility that it might go off in multiple directions" (I am reminded of Hall's injunction that girls "must remain adaptable to a wide range of eventualities"), while "evanescence" is its ephemerality.[129] These attributes can be easily applied to the plastic image, which both sustains and is supported by discursive constructions of girlhood as unfinished, process-oriented, and always changing, somehow both temporary and enduring.

vi.

The way to be a Young-Girl is to be *nothing*.
—Tiqqun, *Preliminary Materials for a Theory of the Young-Girl*[130]

The Girl seems to be portrayed in horror [films], as well as in its critical feminist interpretations, as merely to-be-woman. It may be, however, the Girl's indefinability, her no-thingness, that is the actual lure and promise of free entry. The Girl appears as an empty territory to be occupied either by forces from outside or by the awakening internal forces of female sexuality. Her vulnerability to these possessions may, however, also unveil the threat she poses.
—Taru Elfving, "Haunted: Writing with the Girl"[131]

The plastic image is marked by profound contradictions. On the one hand, it is a figure of resistance. The nimble, labile girl-subject of digital culture can squeeze through the cracks of oppressive power structures. For Gilles Deleuze and Félix Guattari, girls can embody a liberatory process of becoming-woman available to all as a movement between architectures of binary thinking and an escape from the mise-en-scène of the Oedipal family. "Girls do not belong to an age group, sex, order, or kingdom: they slip in everywhere, between orders, acts, ages, sexes; they produce *n* molecular sexes on the line of flight in relation to the dualism machines they cross right through."[132] *A Thousand Plateaus* extends a line of thought from Deleuze's solo-authored *The Logic of Sense*, which submits Carroll's Alice as a figure of becoming. Although the Deleuzian understanding of girlhood is controversial within feminist theory, as it evacuates the girl of her sex and gender specificity in the service of everyone everywhere, feminists such as Catherine Driscoll, Rosi Braidotti, Patty Sotirin, and Elizabeth Grosz have all argued for the political potential of becoming-woman and the micromovements it acknowledges and facilitates.

On the other hand, the plastic image is also a vehicle of hegemony. If girlishness can be a promise, it can also be an imperative. Plasticity can satisfy a neoliberal mandate to respond to market demands and meet the needs of "liquid modernity" by remaining lithe, metamorphic, and responsive.[133] Along these lines is Brian Holmes's damning critique of what he calls the "flexible personality," the embodiment of "a contemporary form of governmentality, an internalized and culturalized pattern of 'soft' coercion, which nonetheless can be directly correlated to the hard data of labor conditions, bureaucratic and police practices, border regimes and military interventions."[134] His version of the plastic image is a response to and a personification of the following socioeconomic changes inaugurated by postindustrialism:

> First, the pressure of a rigid, authoritarian hierarchy is eased, by eliminating the complex middle-management ladder of the Fordist enterprises and opening up shifting, one-to-one connections between network members. Second, spontaneous communication, creativity and relational fluidity can be encouraged in a network as factors of productivity and motivation, thus overcoming the alienation of impersonal, rationalized procedures. Third, extended mobility can be tolerated or even demanded . . . allowing work to be relayed through telecommunications channels. Fourth, the standardization of products that was the visible mark of the individual's alienation under the mass-production regime can be attenuated . . . Fifth, desire can be stimulated and new, rapidly obsolescent products can be created by working directly within the cultural realm as coded by multimedia in particular . . . resolving part of the problem of falling demand for the kinds of long-lasting consumer durables produced by Fordist factories.[135]

Although he doesn't mention gender or age, his "flexible personality" and related "Connectionist man" clearly have girlish attributes.[136] In the case of La Tana di Alice, girlification is the production of a "Connectionist" European identity through the smoothing over of national, cultural, and racial distinctions — undoubtedly in the service of post-Fordist globalization.

On the one hand, on the other hand: the irreconcilable differences between these two hands, the noise they make as they struggle against each other and occasionally embrace, echo the ambivalence of networked culture. Digital tools and global networks do allow for a multiplicity of individual voices to be heard. However, both multiplicity and individuality are appropriated, branded, packaged, and commodified. Sianne Ngai makes a similar point about the dynamics of power in analog animation. While "the excessive energy and metamorphic potential of the animated body make it a potentially subversive or powerful body" (a position she rightly attributes to Eisenstein), the animated body is also pliable, vulnerable "to external manipulation and control."[137] Although the focus of Ngai's article is racialized Asian American and African American characters, her analysis illuminates the contradictions inherent to the animation of other marginalized performing bodies (bodies that had already been put in relation in Eisenstein's discussion of the plasmaticness of African American Snake Dancers and Carroll's Alice).

Philosopher Catherine Malabou explains the possibility and threat of transformation through the distinction she makes between true plasticity and the flexibility that often passes as such. Plasticity "means at once the capacity to *receive form* (clay is called 'plastic,' for example) and the capacity to *give form* (as in the plastic arts or in plastic surgery)" as well as "the capacity to annihilate the very form it is able to receive or create. We should not forget that *plastique* . . . is an explosive substance made of nitroglycerine and nitrocellulose, capable of causing violent explosions."[138] Conversely, "to be flexible is to receive a form or impression, to be able to fold oneself, to take the fold, not to give it. To be docile, to not explode . . . Flexibility is plasticity minus its genius."[139] Malabou separates "simply flexible identity and a truly plastic identity—a distinction resting on a theory of transformation."[140] The former is accommodating, submissive, and, above all, employable, while the latter both accepts and resists outside forces, both gives and takes form, both creates and destroys.

However, I find it hard to root out the entirely enabling plastic images from the field of dangerous ones. As we have already seen with Bellmer's doll photographs, the invocation of girlish transformation can be at once violent and critical of violence. The same Deleuzian concept that problematically appropriates, deracinates, and then empties the girl of her specificity is also a line of flight for certain feminists seeking out dynamic, nonessentialist notions of identity. And Alice's protodigital scalability both traps her within the White Rabbit's home and allows her to take it over, to take up space, to Occupy it. The subsequent three chapters will map these uncomfortable and conflicting forces of girlification in early twenty-first-century digital cultures.

vii.

A theory of cinema is not "about" cinema, but about the concepts that cinema gives rise to and which are themselves related to other concepts corresponding to other practices, the practice of concepts in general having no privilege over others, any more than one object has over others. It is at the level of the interference of many practices that things happen, beings, images, concepts, all the kinds of events. The theory of cinema does not bear on the cinema, but on the concepts of the cinema, which are no less practical, effective or existent than cinema itself.
—Gilles Deleuze, *Cinema 2: The Time Image*[141]

I imagine that right now you're feeling a bit like Alice, tumbling down the rabbit hole.—Morpheus to Neo, *The Matrix*[142]

Chapters 2, 3, and 4 offer a theory of girlishness. This is ultimately not a theory "about girlishness" (and even less a theory about girls), but about the concepts that feminine adolescence gives rise to and which are themselves related to other concepts corresponding to other practices. Girlhood is itself a practice, a performative process that is now, I argue, inextricable from ways of understanding and deploying images in the Information Age. Girlhood is related to the concept of plasticity corresponding to practices of animation, advertising, protodigital photography, social media, web design, Internet art, and digital cinema, among others. It is at the level of the interference of many practices that things happen; beings, images, and concepts are brought to life; and plastic girlhood becomes an anxious and exhilarating global event. The films and artworks I discuss in the remainder of this book are not illustrations of the theories of Deleuze, McLuhan, or Hall, but are themselves generative theories in their own right, as the many Alices I follow bring a mode of girlishness into being that anticipates and, in part, produces digital plasticity. The very transmedia nature of girlishness requires an approach that positions textual and visual objects of study on a single plane and then traverses the borders between them.

Each of the following chapters addresses a single but unusually expansive work—Pierre Huyghe and Philippe Parreno's *No Ghost Just a Shell* (1999–2002), Martine Neddam's Mouchette.org and its satellite websites (1996–present), and Satoshi Kon's *Paprika* (2006). They all animate their own versions of a metamorphic and mobile, malleable and vulnerable, vacuous and connective Alice in Wonderland. In fact, all are centrally, even obsessively concerned with the production of a plastic girl — Annlee in *No Ghost Just a Shell*, Mouchette on the websites that bear her name, and Paprika in Kon's animated film — gesturing toward the increasing influence of character design in the Information Age.

Linking these figures to related examples within fine art and popular culture, my argument moves from high-profile media objects (*Paprika*, for example, is one of the most widely known anime films of the last fifteen years) to works produced by emerging artists (such as art student Emilie Gervais's *Parked Domain Girl Tombstone*), from celebrities (Hello Kitty) to minor players (Tasha from the computer game *Leisure Suit Larry in the Land of the Lounge Lizards*). Our flattened media universe, in which a serious news report is given the same amount of space in the Google search results list as a personal blog post, requires scholars to reassess traditional hierarchies of academic research. I believe that theories of the digital image must address malleability across scale, from the Photoshopping of an individual's Facebook profile pic to the softening of conventional boundaries between fine art and popular media.

Even scholarship that claims to account for "the image" can have a surprisingly narrow focus. To wit, David Joselit's *After Art* (2012) offers a theory of the contemporary image in the age of networked globalization, but makes little direct reference to mass media. Despite provocative claims like "The more points of contact an image is able to establish, the greater its power will be" and an argument that considers the status of the image writ large, he remains staunchly committed to the fine arts.[143] His book is otherwise compelling, but in its disregard for the kind of image that, by definition, maximizes "points of contact," it is at best a missed opportunity, at worst an obfuscation of the intellectual provenance of some of his argument's key vocabulary. *After Art* purports to understand value as "saturation through mass circulation" only to betray its own terms.[144]

Annlee and Mouchette, who cross the line between mass media and fine art, and Paprika, who resides squarely in the former, are "after art" stars who have attained saturation through mass circulation and generated points of contact through the World Wide Web. I have chosen these girls and the rest of the book's cast of characters for their power to reorient late-modern sociotechnological processes of image circulation, consumption, postproduction, participation, globalization, and remediation toward girlishness. Or rather, they reveal the preexisting alignment of these practices toward a constellation of attributes that are often too small or too diffuse for us to notice. Girlish plasticity is not only a telescope, but also a microscope — either way, an optic for viewing the most fundamental nano- and macro-aesthetics of the digital image.

two. Sharing Girlishness

She was Lo, plain Lo, in the morning, standing four feet ten in one sock. She was Lola in slacks. She was Dolly at school. She was Dolores on the dotted line. But in my arms she was always Lolita. — Vladimir Nabokov, *Lolita*[1]

Work with her, in a real story, translate her capabilities into psychological traits, lend her a character, a text, a denunciation and address to the Court a trial in her defense. Do all that you can so that this character lives different stories and experiences. So that she can act as a sign, as a live logo.
— Pierre Huyghe and Philippe Parreno, instructions for artists participating in *No Ghost Just a Shell*[2]

i.

In 1999, French artists Pierre Huyghe and Philippe Parreno bought the rights to a character created by K Works, a Japanese company that produced figures used in comic books, cartoons, video games, and advertising. The image was of a girl with purple hair and a mournful expression, one of those cute, saucer-eyed anime figures dominating Japanese visual culture as it is exported to the West. Her name was Annlee (plate 6).

Out of all the images in the K Works catalog, Annlee was chosen because she was inexpensive and minimal, without the complex characteristics given more pricey figures. According to the artists, her spare profile — her lack of heft — marked her for death in the worlds of anime and manga. As Annlee explains in one of Parreno's contributions to the project, the animated video *Anywhere out of the World* (2000), "Some other characters had the possibility of becoming a hero. They had a long psychological description, a personal history, material to produce a narration ... Designed to join any kind of story, but with no chance to survive any of them ... Drop dead in a comic book."[3] The artists label their own practice as one of emancipation and equate commercial media with alienation and murder.

Huyghe and Parreno shared the Annlee digital file with an international group of artists, each of whom created a work featuring the character. The resulting collection, *No Ghost Just a Shell,* includes more than twenty-five different pieces in several different media — from animated digital videos (the keystones of the project) to sculpture, painting, sound art, and graphic design. Annlee's identity in this series is profoundly variable. She appears in 3-D and in 2-D; as a still image and as a moving animation; and as a graphic line drawing on video, on paper, and in neon light. She has different hair colors and outfits, is voiced by different people (sometimes within the same video), is multiplied and dispersed. Even her name has alternate spellings. She is identified as Annlee, Ann Lee, AnneLee, and Anlee (like Lena/Lenna, the test image and *Playboy* centerfold I discuss in the previous chapter). "Spell it however you want! It doesn't matter," she tells us in *Anywhere out of the World*.[4]

Annlee changed form as she traveled through the transnational network of exchange. A plastic image, she was appropriated, altered, and inserted into a dizzying array of artworks and narratives. For example, she stars in Huyghe's *One Million Kingdoms* (2001), a short animation depicting her wandering amid a desolate lunar landscape (plate 7). This digital video combines text from Jules Verne's *Journey to the Center of the Earth* (1864) with an image of our heroine reproduced as a glowing blue outline. In Rirkrit Tiravanija's eight-and-a-half-hour animated video *Ghost Reader C.H.* (2002), Annlee is rendered as a blocky 3-D figure reading the entire text of Philip K. Dick's novel *Do Androids Dream of Electric Sheep*? (1968). Dominique Gonzalez-Foerster's *Annlee in Anzen Zone* (2000) features cloned Annlees — looking similar to the one in *Ghost Reader C.H.*, but with more refined contours — speaking in English and Japanese about an upcoming invasion. Joe Scanlan's *DIY, or How to Kill Yourself Anywhere in the World for under $399* (2002) presents Annlee as a black line drawing in a cheeky instruction manual illustrating how to build a coffin out of repurposed IKEA furniture. And for *Smile Without a Cat (A Celebration of Annlee's Vanishing),* a live event marking the end of the *No Ghost Just a Shell* collaboration, fireworks forming a blazing outline of Annlee's face were set off at 2002's Art Basel Miami Beach fair (plate 8).

Despite the instability of her identity, Annlee binds together both the sprawling *No Ghost Just a Shell* series and the group of participants. She is a shared symbol of togetherness for the artists involved in the project, demonstrating the image's "capacity for 'linkage' [*reliance*], to use Michel Maffesoli's term: flags, logos, icons and signs all produce empathy and sharing, and generate *links*."[5] Indeed, the plastic image of Annlee has considerable power to generate links, expanding the frame of *No Ghost Just a Shell* to include the relations between the participating artists, the encounters between viewers and the artworks, the interactions between artworks (as well as between different versions of Annlee/AnneLee/Anlee/Ann Lee), and — most important for my purposes — the discourse surrounding the artworks.[6]

Surprisingly, critics play along with Huyghe and Parreno by continuing their efforts to bring the Annlee image to life. In other words, Annlee is further animated by *No Ghost Just a Shell*'s discursive field. Descriptions of Huyghe and Parreno's act of purchasing the image portray the artists as fairy-tale heroes liberating a princess from the market economy, as husbands procuring a foreign wife (a review by Marcia Tanner is entitled "Requiem for a Mail Order Bride"), and as pedophilic adoptive fathers allowing their artist friends "to have their way with her."[7] There are two key aspects of this widespread personification of the Annlee image: first, that the image is vivified, and second, that the image is girled. If the subject of *No Ghost Just a Shell* is not Annlee but a network of links between artists, artworks, and viewers, then this dispersed and makeshift collective holds together through the shared effort of bringing an image to life as a girl who is either rescued or corrupted by Huyghe and Parreno.

No Ghost Just a Shell — a conceptual performance, an experiment in collaborative vivification, a set of artworks, a series of exhibitions, a collective of artists, an art historical object of study, and a network of shifting relations — is a privileged site in which to examine practices of girlification in contemporary image culture. Indeed, it enacts girlhood in several interrelated ways. As a mail order bride or abused daughter, Annlee corresponds to an image of girlhood that dominates popular discourse: the girl as innocent victim. As a product of globalized commercial culture, she functions as a feminized commodity adopted, transformed, and allegedly elevated by a masculinized artistic gesture. And as a digital image that changes form as it changes hands, Annlee offers girlhood as *No Ghost Just a Shell*'s formal condition. Annlee/Anlee/Ann Lee/AnneLee, with the plasticity of her identity, functions as a flag, logo, icon, or sign of a multimedia art project with variable parameters. The unfixed and unstable nature of the project finds its conceptual emblem in the supposedly unfixed and unstable nature of girls. In this sense, *No Ghost Just a Shell* coheres not despite, but because of, the mutability of Ann Lee's appearance.

Huyghe and Parreno's desire to create a polymorphous artwork depends on understandings of girls as infinitely plastic images and of girlhood as a state of potentiality. As I argue throughout this book, girls are interpellated as digital/protodigital images — that is, they are performatively produced as to-be-looked-at, should-be-shared, and can-be-altered. Aptly, Huyghe explains Annlee's appeal using terms that resonate with later chapters of this book: "She's a polyphonic character . . . What's interesting about this manga figure is that it's a way to tell a story. A sexual story? You can use her. A dark story? You can use her. A nice story? You can use this character. She's almost like a tool."[8] Yet as this quote suggests, Huyghe seems unsure as to the gender of Annlee, who is alternately a "she," an "it," and (almost) a "tool." Is Annlee an object, or is she just objectified? How does Huyghe's equivocation fit into my larger claims regarding the works' performance of girlhood and animation of the image?

Emphasizing "it" over "she," art historian Tom McDonough sidesteps the personification of Annlee. Despite his admission that Huyghe and Parreno have participated in the shared performance of anthropomorphization, McDonough maintains that, ultimately, there is no ghost in this shell — as the title attests — no "flicker of humanity."[9] Indeed, McDonough writes (in relation to the animated video *Two Minutes Out of Time*), "Huyghe seems determined to remove the image from any identification; it remains, in some sense, beyond recognition."[10] For McDonough, Annlee is not human; she is a sign of the artists' desire to move beyond identity, subjectivity, and — we are to assume — gender.

McDonough supports his claims with quotations from the artists that counter the "naïve set of assumptions" plaguing the project.[11] For example, Huyghe chastises a critic with the imperative "Don't make it romantic" in response to the latter's description of Annlee as a sexualized fantasy.[12] Buttressing McDonough's assertions (but not mentioned in his article) is Huyghe's apparently definitive statement "It isn't a 'she,' it's an 'it.'"[13] McDonough's totalizing argument for the project's radical postsubjectivity denies any kind of identification of or with the image.

There is more to this story, however. What McDonough doesn't tell us is that "Don't make it romantic" is followed, in the original source, by "'She's just a virgin,' [Huyghe] continued, still laughing. 'We bought a virgin.'"[14] This omission supports McDonough's gender-neutral reading of the project. Ultimately, I maintain, Huyghe and Parreno are not a pair of myth busters challenging viewers' and critics' naive desires to vivify an image. Instead, they give us a far more ambivalent portrait of Annlee as a thing, a digital file, an "it"—for McDonough, a postsubject—as well as a girl, a virgin, a "she."

In opposition to McDonough's insistence on postsubjectivity, my own reading of *No Ghost Just a Shell* explores the gendering of Annlee despite—or, rather, because of—the artists' intermittent insistence on her thingness. I argue that Huyghe's, Parreno's, and McDonough's failed attempts at neutering the image are made possible by the pervasiveness of girlhood in image culture. Indeed, the hypervisibility of girls in the media can obstruct our view of the process of gendering and aging plastic images. Girls are everywhere but nowhere; fetishized but neglected; in our faces yet disappearing from view. Furthermore, what McDonough considers to be a postsubject I recognize as a plastic girl, an icon of gendered transformation at the center of *Alice in Wonderland* as well as Mouchette.org/.net and *Paprika*, the subjects of subsequent chapters. Annlee is not only a girl, but also a paradigm of digital girlishness, demonstrating all of the key functions of the plastic image.

As Disney's *Alice's Wonderland* and *Alice in Wonderland* are part of a wider discursive network that gives meaning to girlishness as an image-bound phenomenon, *No Ghost Just a Shell* is embedded within a matrix of relations that includes the performance of girlhood in both fine art and the mass media. What if Annlee's origin in comic books, advertising, and animation is used not to support fairy-tale narratives of emancipation from the commercial image economy, but instead to help us better understand the function of girls within *No Ghost Just a Shell*'s expansive network of linkages?

This methodology is appropriate in light of Nicolas Bourriaud's articulation of postproduction, a set of strategies associated with Huyghe, Parreno, and other artists who came to prominence in the 1990s. Postproduction is a term usually applied to commercial filmmaking, television, audio, and photography; it refers to the modification of imagery or sound after the initial recording, now accomplished through various software packages.

In Bourriaud's book, the DJ who samples and remixes tracks, the computer programmer, and the web surfer are all conceptual personas for postproduction artists who appropriate and transform source material (as did the participants in *No Ghost Just a Shell*).[15] As Hal Foster summarizes, the rise of postproduction techniques in art represents a move from the age of production to the age of information, with the relative ease of appropriation, manipulation, alteration, and dissemination that data facilitates.[16]

With this in mind, we can consider Annlee to be an incarnation of the kind of digital mutability that both contributes to and emerges from strategies of postproduction. The sharing, exporting, and reformatting of Annlee put *No Ghost Just a Shell* alongside Douglas Gordon's *24 Hour Psycho* (1993) and Pierre Huyghe's *The Third Memory* (2000), both of which are presented in *Postproduction* as paradigmatic examples of art generated through the manipulation of preexisting images.[17] As I explain in chapter 1, Girl Talk is not only the moniker of postproduction musician/DJ Gregg Gillis. It is also a mode of engagement between consumers/producers and the plastic images and sounds that circulate across global networks.

Annlee's girlhood is both rarely forgotten and rarely analyzed. What follows is a series of answers to this simple question: Why is Annlee a girl? The multiple rejoinders I provide neither lionize the project nor tear it apart. Instead, I seek to understand the radically ambivalent discourses of girlphobia and girlphilia that uneasily come together to create *No Ghost Just a Shell.*

This chapter also addresses a cluster of related questions concerning *No Ghost Just a Shell*'s mechanisms of girlification. How does the project perform girlishness through a field of actions including more than twenty-five different individual works and several different exhibitions? How do *No Ghost Just a Shell*'s strategies of postproduction produce the virginity and plasticity of its source? And, more broadly, how can *No Ghost Just a Shell* help us trace a relationship between girls, media, commodification, and the image — from the turn-of-the-twentieth-century Tiller Girls to the turn-of-the-twenty-first-century Annlees? Annlee reminds us of the long-standing feminization and brutal infantilization of girl-products. At the same time, her girlish figure embodies new beliefs in the possibility of reforming girl-products, marking key changes in our attitudes toward the buying of virgin images in the twenty-first century.

ii.

To inhabit the child, then, is to inhabit the condition, once again, of possibility itself; and to inhabit the condition of possibility is to become a child.
—Claudia Castañeda, *Figurations*[18]

Nicolas Bourriaud uses the title of one of Huyghe's early pieces—*Chantier permanent (A permanent construction site)*—as an apt descriptor of the artist's broad oeuvre. For Huyghe, form is "never considered as a finished product, but rather as the fruit of a slow elaboration."[19] In the case of *No Ghost Just a Shell*, the form under elaboration is Annlee, whose "ambiguous status as a girl without qualities" allows her to be resculpted by the desires of the participating artists.[20] *No Ghost Just a Shell* animates the girl's formal condition. What is brought forth is the form of the girl as both action and object, as a process of elaboration and as the fruit of that (e)labor.

Huyghe and Parreno didn't just purchase a product—a commercial form—and reformat her. They and their collaborators *girled* that product through a collective performance of appropriation, dissemination, and alteration. Consequently, the girled image is produced as that which needs to be bought, rescued, possessed, exported, passed around, and transformed. *No Ghost Just a Shell* insists that this engendering of images and girls is both a practice and a group project(ion). Annlee never just is; she becomes, through collective labor. She is a body under construction.

The plasticity of the Annlee image works in concert with ideas concerning the developmental incompleteness of young people. As Claudia Castañeda argues about children of both genders (although, I would say, this is especially applicable to girls), "The mutability of the child figured as a body in process makes it eminently appropriable; not yet fully formed, it has no prior being that must be displaced and then re-placed."[21] For Castañeda, the child's "embodied transformation" is produced through a network of power relations that confer value on children and their capacity for metamorphosis, emptying youth of their historical, geographic, racial, cultural, and material specificity.[22]

Although Castañeda's case studies are significantly different from my own—developmental science, adoption, child organ theft—her approach recognizes the entwining of children and images. For example, her examination of transnational and transracial adoption incorporates a reading of the infamous cover of an immigration-themed issue of *Time* magazine from 1993, discussed briefly in the previous chapter.[23] The young woman pictured on the cover was created by morphing together images of people with different national and racial backgrounds using the software Morph 2.0. This form of virtual miscegenation imagines reproduction as the merging of surface characteristics and the creation of progeny without deeply embodied racial identity, national affiliation, or "historical contingency."[24] In a now familiar move of vivification, several *Time* employees admit to falling in love with their creation, reminding us that images are both generated and animated by an unrelenting desire to make it romantic.[25]

Time's cover model is a data body vivified through a series of metamorphic operations aligning software with reproduction, malleability, and flexible hybridity. Instead of a provocative challenge to white America's configuration of the Nation, America's techno-daughter is an embodiment of fantasies of the unraced, deracinated, girled plastic subject; like an anime girl, she appears to be of more than one race and no particular race all at once. She functions as both an effect and confirmation of soft(ware) power.

The promise of "embodied transformation" is the ontological condition of the girl fashion model, whose exportability is based on the plasticity of her image. In the fashion and beauty industries, girls are valued for their potentiality, the ease with which they can be made and remade into different 2-D images. Software enhances the bodily resculpting these girls do on their own, morphing—or perhaps I should say *girling*—the contours of their figures. Annlee is like a model (a simile made literal in the animated video *Anywhere out of the World*, in which the voice-over actress is a model named Daniela). Her undeveloped figure leaves her open for transnational appropriation by Huyghe and Parreno, who doubly girl the image by exporting and reformatting her.

No Ghost Just a Shell gives us both a paradigmatic girl image and a paradigmatic digital image. Indeed, it demonstrates the imbrication of girlhood and digital media as they are discursively constructed. As chapter 1 articulates in detail, through examples from both popular image culture and digital image theory, both girls and digital objects are deemed shareable and manipulable. When Lev Manovich writes that "computer representation makes every image inherently mutable — creating signs that are no longer just mobile but also forever modifiable," he could just as well be referencing the bodies of girls, whose cultural and economic exchange value lies in the assumed plasticity of their contours.[26] Girls (such as teenyboppers, starlets, reality TV girls, pageant girls, muses, songstresses, supermodels, makeover contestants, America's sweethearts, music video girls, and divas) thrive in the image economy because of digital mutability; conversely, digital mutability has been incorporated into our own desires for a never-ending youthful plasticity. When a participant on a makeover show says of her own postbaby body, "It's a picture, it's an image, everything can be changed," she goes one step further than Andre Agassi, who delivered the now infamous tagline "Image is everything" in a 1990 commercial for Canon cameras.[27] The young mother reminds us that in our contemporary mediascape, Agassi's statement is more meaningful in the reverse: everything is image, which means that everything can be changed. Manovich's principle of variability is really an embrace of girlishness as the formal condition of digital identity.

As variable creatures, both girls and digital images manifest the "aesthetic of unfinish," Peter Lunenfeld's useful phrase for the aesthetic of digital media.[28] Digital forms, Lunenfeld argues, are fundamentally process-oriented and incomplete, "not only allow[ing] for constant incremental changes but demand[ing] them."[29] He offers the aesthetic of unfinish as a lens through which to view digital architecture, websites, software, hypertext fiction, and transmedia film, all of which are objects-under-construction by creators and often users. In Lunenfeld's writing, "unfinished" has a number of connotations that resonate with familiar understandings of childhood and, especially, girlhood — such as unbuilt, unrealized, unresolved, incomplete.

As Bourriaud has his DJs and web surfers (conceptual figures who embody his central concepts), Lunenfeld has James Dean and Marilyn Monroe, both of whom serve as personifications of digital aesthetics. Their unfinished lives — tragically cut short, underfulfilling our hopes for their success — are paradoxically more successful in their eroticized incompleteness. After all, "the unfinished work or person allows us to read our own desires into a not yet fully formed object — opening up more space for pleasure and identification than any 'complete' work or person can ever offer."[30] Although Lunenfeld does not examine the gendering or aging of his conceptual personae, both James Dean and, more obviously, Marilyn Monroe have been girled by popular media. Dean, in particular, is now often remembered as an icon of virile masculinity; however, he was, in his time, strongly associated with adolescence, androgyny, and an intensely emotional vulnerability. A 2005 newspaper article by Germaine Greer describes him in these terms:

> Everyone who knew James Byron Dean in real life called him Jimmy; the diminutive was unavoidable because he was small in stature. He looked junior. He would be referred to as "boy" and "kid" all his short life. At a crucial point he realised that if he grew himself a tall quiff, left the glasses off and pouted a little, he could look really tasty.[31]

While it is beyond the scope of this chapter to perform an extended analysis of Dean's powerfully queer adolescent persona, I do want to return to a point I made in chapter 1: as performatively produced, girlishness is dislodged from its sexed characteristics. Digital and protodigital girlhood is a mobile phenomenon that can be applied to men and women, adults and children, even nonhuman animals and objects. Of course, "girlie" is also used as a slur against men who are not convincingly masculine; taking the word from a *Saturday Night Live* sketch roundly mocking his overinflated musculature, Arnold Schwarzenegger repeatedly called those with opposing political views "girlie men." For Dean, his "tasty," pouty, "junior" girlishness instead worked in his favor, as it made him better able to be shared by others and reformatted to fit the desires of both men and women. Forever a boy, and a girlie boy at that, Dean performs the "condition of possibility" that Castañeda finds in discursive constructions of children.

Let us focus in on the aesthetic of unfinish, crucial to understanding Annlee (and, by extension, the girl) as a media object. One of Lunenfeld's central points — that the unfinished image facilitates greater participation — was made much earlier by McLuhan in *Understanding Media: The Extensions of Man* (1964). Published thirteen years after his first book (*The Mechanical Bride*), *Understanding Media* offers a more nuanced account of the effects of media forms and technologies on consuming subjects. McLuhan's boldly antithetical take on mechanical brides gives way to a kind of thoughtful ambivalence, and his consistent interest in the gendering of images and the emasculation of viewers becomes more of an intermittent concern. Nonetheless, references to gender do appear in explanations of some of his most influential concepts. Take, for example, McLuhan's articulation of the difference between cool and hot media. Relying on one of his characteristic metaphors as a method of argumentation, McLuhan turns to the cliché "Men seldom make passes at girls who wear glasses" as an explanation of hot media — or media that invite passive spectatorship. He adds: "Glasses intensify the outward-going vision, and fill in the feminine image exceedingly, Marion the Librarian notwithstanding. Dark glasses, on the other hand, create the inscrutable and inaccessible image that invites a great deal of participation and completion."[32] Continuing the tradition of feminizing the object of the gaze, McLuhan figures the viewing experience of cool media — which include cartoons, television, and other media lacking the high-definition information of live-action cinematic images and, less obviously, radio — as heterosexual male attraction to girls in sunglasses. The alluring, dark void of their orbits encourages viewers to fill in the blanks and complete the picture. Long before Lunenfeld invokes icons Marilyn Monroe and James Dean as embodiments of digital media's unfinished aesthetic, McLuhan explains his own theory of interactivity through the distinction between two archetypes of girlhood (between, we could say, the bookworm and the starlet).

I want to return to Marilyn Monroe, and not only because of the many photographs of her sporting cool, classic Wayfarer sunglasses. Despite Monroe's womanly curves, she is, in some ways, a surprisingly appropriate companion for Annlee. Monroe cultivated a girlish performance persona well into her thirties. Her light blonde hair (associated, especially at the time, with childhood), cartoonish high-pitched voice, and doe-eyed innocence counterbalance the knowing sexiness of her attire and come-hither gait. Branded into public consciousness through numerous movies, news articles, public appearances, magazine spreads, and media scandals (transmedia avant la lettre), the Monroe persona is both remarkably consistent and surprisingly plastic. Monroe, like Annlee, is a "polyphonic character" able to be inserted into a multitude of stories.[33] Indeed, the popularity of the dumb blonde archetype she perfected is due to its transmedial and transhistorical adaptability; Monroe is an evacuated media object able to be filled with multiple desires (and even performed by multiple people). One of Monroe's most iconic lead roles was that of a model in *The Seven Year Itch* (1955). Her character in that film is known only as "The Girl."

I am not arguing that Monroe really was dumb — quite the contrary — or that women in sunglasses do not give knowing glances behind the darkness of their shades. Instead, I want to do two things: establish the girlish prehistory of digital aesthetics and highlight the aesthetic of unfinish inherent to enactments of girlhood in visual culture. If women are supposed to be finished paintings, then girls are believed to be sketches — that is, cartoons (to invoke a lesser-known meaning of the word as a rough drawing created in preparation for making a fresco). The goal of the performance of girlhood in media is to extend, infinitely and forever, the cartoon in time and space. Monroe's early death allows her animated image to continue to circulate without the encumbrance of old age. She will always be the cartoonish "Girl" model — or showgirl or prostitute or girlfriend, depending on the movie — both nothing and everything to everyone. She will always be seen as a "female tabula rasa," an "empty vessel ripe for exploitation," and (if not a virgin) an abused child bride.[34] Her image is — to quote Brian Eno's explanation of both aesthetics and identity in the age of transmedia — "permanently unfinished."[35]

Annlee is a permanently unfinished cartoon for our digital age (indeed, she is much less complex and data-rich than the Monroe image). Her unfinished aesthetic is apparent in the overall structure of a project based on the infinite potential of a supposedly empty image. Additionally, the aesthetic of unfinish is manifested by the project's multiform exhibition history. Each showing of *No Ghost Just a Shell* is significantly different, and some of the works have been exhibited independently of the rest of the collection. As Pierre Huyghe insists, "No exhibition is a final representation." Moreover, "if you consider each appearance of the project as a singular object, then we're talking about an object that's always in the process of multiplying, of morphing."[36] Stefan Kalmar adds, using the now familiar language of animation, that "Annlee is traveling, adapting herself to four different venues (Zurich, Cambridge, San Francisco and Eindhoven)," the four sites of the exhibition at the time of the catalog's publication.[37]

The aesthetic of unfinish also appears within the frame of individual works. For example, Huyghe's beautiful digital animation *One Million Kingdoms* shows Annlee — here, a glowing blue outline suggestive of the Annlee neon sculpture (also part of the exhibition) — walking alone through an unstable topography (see plate 7). The bluffs and rocky outcrops of her environment rise and fall as she approaches, suggesting the real-time rendering of environments in 3-D computer games. The considerable negative space of Annlee's body draws attention to the shape-shifting extraterrestrial landscape; indeed, the outline of her form frames the movement of the unfinished world around her and inside her. This planet — or, better, this moon — is a construction site incorporated into the void of Annlee's body. At the end of the video, the landscape suddenly fragments in slow motion. Bits and pieces move steadily into the darkness along angular trajectories. The pixilation of the moon cuts short its "life" yet ensures its never-ending expansion. The effect is airless, forlorn, deeply melancholic.

One Million Kingdoms does not appear to be about girlhood — at least, not explicitly. Most obvious is the video's linkage of Jules Verne's *Journey to the Center of the Earth* to Neil Armstrong's moonwalk. Annlee's voice is an altered version of Armstrong's transmission from the moon; she speaks some of Armstrong's words as well as appropriated text from Verne's novel. The intersection of Armstrong and Verne's narrator is Iceland: characters in *Journey to the Center of the Earth* began their descent in an Icelandic crater, while astronauts training for the lunar landing explored sites in Iceland in order to familiarize themselves with volcanic terrain. The question is, what does Annlee have to do with these rehearsals for extraterrestrial exploration?

Like a ventriloquist's dummy (which, not incidentally, almost always takes the form of a child), Annlee is offered as a shell that can be filled with any number of voices originating from outside of her body. In this case, the voice in question is a digitally manipulated version of that of Armstrong; Annlee's small blue line of a mouth moves in phonemic lockstep. As the film proceeds, words taken from the lunar transmission (including one half of the famous "One small step for a man" statement) give way to an English translation of parts of chapters 17 and 23 of Verne's novel:

> I had not yet looked down into the bottomless pit which I was about to plunge into. I was ashamed to step away. Now the time had come. I land over a projecting rock and look down. My hair stood on end. The fascination of the void took hold of me. I felt my center of gravity moving and vertigo rising to my head like intoxication. There is nothing more powerful than this attraction of the abyss. Nothing, with the possible exception

> of the state of weightlessness. For a moment I was afraid that their words might be my own brought back by an echo. I listened once more, and this time I clearly heard my name thrown through space.[38]

Spaces defined by their emptiness, voids are radically unfinished; here the bottomless pit serves as an "inscrutable and inaccessible image that invites a great deal of participation and completion" (as McLuhan would say). Both Iceland's volcanic terrain and the moon have served, historically, as "cool," low-def sites on which non-Icelanders and earth dwellers project their fantasies and fictions of nothingness/everythingness.

One Million Kingdoms suggests that desire for Annlee is a kind of sublime and anxious attraction to the abyss. I am reminded of Lunenfeld's review of a work by artist Ray Johnson: "Shirley Temple, 1958, is a tiny piece — just the little moppet's silhouette inside a homespun, cardboard collage — but it demonstrates how Hollywood invades our imaginary, forcing us to fill in even the emptiest of signifiers."[39] *No Ghost Just a Shell* presents Annlee as a chasm, an empty signifier, a nothing filled in through the combined desires of a network of artists.

However, according to the logic of *No Ghost Just a Shell*, we are not *forced* to fill Annlee with meaning, but *compelled* by her vacuity. Annlee is a representative less of lack than of the connectivity that emptiness facilitates. As such, she can be productively compared to the Barbie doll as understood by Catherine Driscoll. "Anything that looks like lack on Barbie is really the potential for further attachment," Driscoll argues, partially reclaiming Barbie as an icon of the aesthetic of unfinish. "Barbie functions as a recording surface for desires and identities."[40] Despite the stiffness of her joints, the hard plastic of her flesh, and the almost unchanging proportions of her figure throughout the decades, Barbie can be said to embody mutability. As Don Clarke, former vice president of design and development for Mattel, insists, "Barbie *is* change . . . There was Barbie the model (1959), Barbie the career girl (1963), Barbie the surgeon (1973), and Barbie the aerobics instructor (1984)."[41] Barbie's emptiness allows her to network with any number of narratives, both those officially sanctioned by Mattel and those invented by children through resistant play. Caught in the "permanent state of becoming-a-woman" (that is, permanently unfinished), Barbie materializes both the horrifying and the enabling aspects of our attraction to the abyss.[42]

Importantly, Annlee — like Iceland, the moon, Barbie, Shirley Temple, and images of girls more generally — is not *truly* a void, no matter how connective she is. As Castañeda claims, children and images of children are emptied of meaning in order to remake them as vessels for projection. Their capacity to embody unfettered potentiality masks the material relations and individual histories that bring each child and image into being. Annlee, like all images, has a past, present, and future, a set of conditions from which she emerged, a network of relations (corporate, media, fiscal, art historical) that allow her to take shape. Plastic images in particular are perceived to be more agreeable to evacuation, appropriation, dissemination, and manipulation than others. When Huyghe, Parreno, and others deny the importance of gender to the project while avowing Annlee's hollowness and "virginity," they paradoxically girl the image by insisting that she is not a girl. Annlee's availability, her unmistakable girlishness, appears to empty her of everything, including the very age and gender that produce her global availability.

As Web 2.0 facilitates the enabling aspects of girlish connectivity, it also sheds light on the troubling voidance of the self that connectivity all too often entails. Barbie can be everything because she is nothing (or, rather, the material somethings of the doll's life history are concealed). Iceland is the moon only to those people who don't live there. And Annlee implores the participating artists to complete her only after being dispossessed of the ghost in her shell.

iii.

The beautifully designed surface of the commodity becomes its package, not the simple wrapping for protection during transportation, but its real countenance, like the fairytale princess who is transformed through her feathered costume in which she seeks her fortune in the marketplace.
— Wolfgang Haug, *Critique of Commodity Aesthetics*[43]

Who is to say where packaging begins and ends in a medium in which everything is composed of the same streams of data — regardless of whether the information is textual, visual, aural, static, or dynamic?
— Peter Lunenfeld, "Unfinished Business"[44]

Parreno's animated video *Anywhere out of the World* begins with a voice-over and an empty, deep blue screen. A feminine voice introduces herself. "My name is Annlee! Annlee! You can spell it however you want! It doesn't matter! No it does not. I was bought for 46000 Yen," she says casually. Then an image of Annlee emerges out of the lonely blue depths, showing us the girlish figure and dark gray eyes attached to the voice we have been hearing. She explains the conceptual gesture that initiates *No Ghost Just a Shell.* "I/am/a product," she asserts with conviction, "a product freed from the market place I was supposed to fill. Drop dead in a comic book."[45] Her manner is direct but not overbearing, open but not chummy. At the end of the video, the voice — identified as that of a model named Daniela — tells us that she, Annlee, "belong[s] to whom ever is able to fill me with any kind of imaginary material [. . .] I am no ghost, just a shell."[46]

Parreno's video familiarizes viewers with key aspects of *No Ghost Just a Shell.* Confirming my explanation of Annlee from the previous section, it presents the original Annlee as a connective void filled by the imaginations and aesthetic sensibilities of the artists participating in the project. The first Annlee — the 2-D digital image of an anime girl with sad, limpid eyes, pointed ears, and a tiny, straight line of a mouth — is described by critics as a "null set," "a nondescript, expendable, empty vessel ripe for exploitation," and "an anonymous figure that could be adapted to any storyline" (see plate 6).[47] Second, we have Huyghe and Parreno's alleged emancipation of Annlee, who was "condemned to death by commerce, given life by art" and "free[d] . . . from the fiction market."[48] Once supposedly liberated from her Japanese prison and imported to the West (a troubling Eurocentric gesture that will be discussed later), Annlee was distributed as a digital file among a select group of artists, several of whom are associated with postproduction strategies of art making and most of whom are European. The anime girl gets a "cosmetic makeover," "an identity, multiple identities, in fact, and a voice, several voices."[49] She appears in paintings by American artist Richard Phillips and French artist/businessman Henri Barande, wallpaper by the French design firm M/M, illustrations by American artist Joe Scanlan, a magazine by French artist Lili Fleury, and nine different animated videos, including one by Buenos Aires–born, New York–based Rirkrit Tiravanija. She is given an LP, some small, white, blobby sculptures, and a coffin. She is also the face of several posters by M/M. One cleverly replaces the word

"Shell" in the project's title with the logo for the multinational corporation Royal Dutch Shell (Shell oil); a large grayscale version of the original Annlee, her faced smudged as if yanked out of an old large-format photocopier, is positioned to the right of the red and yellow shell graphic. Each artist makes Annlee his or her own.

Whether Annlee is emancipated from the Japanese image economy or from the fiction market more generally, the discourse surrounding and producing *No Ghost Just a Shell* identifies her as both a commodity and a victim. A linchpin of the whole project, *Anywhere out of the World* portrays Huyghe and Parreno's artistic gesture as conferring a voice, both literally and figuratively, to an oppressed image and giving her a much-needed additional dimension — once again, both literally and figuratively — by transforming the purchased 2-D image into a moving 3-D CGI animation. Annlee's capacity for survival is dependent on the will of her artist-saviors and the digital plasticity of her image.

At a key moment in the video, she holds up a piece of paper with a picture of her original face. "Look! That's how I used to be! And this is how I look now," she says, shoving the image of her former self in front of the virtual camera (the magnitude of this gesture is underscored by Richard Phillips's painting *Annlee*, a rendering in oil of a still from this moment in the video).[50] The Annlee bought from K Works was cuter, a little sadder, and definitely 2-D, a contrast to the 3-D CGI fingers holding the picture. We see some surprising colored circles — faux lens flare — over the older image, as if it were an analog photograph from a family album and not a digital image purchased from a catalog. By simulating the effect of scattered light on an analog camera lens, Parreno gives a cheeky authenticity to the affect displayed by the dejected girl in the "photo." Huyghe discloses that "it was that feeling, that melancholy in the character that made us choose this image rather than another."[51]

The artist's description of Annlee as melancholic follows a question from interviewer Stefan Kalmar.

> Kalmar: "How far do you think about 'it' as a gendered sign? Would you agree that Annlee is interchangeable, a kind of trans-sexual — or a shell, for projection?"
>
> Huyghe: "It isn't a 'she,' it's an 'it.' It was that feeling, that melancholy in the character that made us choose this image rather than another."[52]

We have heard the statement before: Annlee is an "it," not a she. Annlee is an image. A product. The simplicity of this statement, its comforting commonsensicality, masks the complexity of the rhetorical mechanisms that girl both images and products. Despite Huyghe's, Parreno's, and others' statements to the contrary, *No Ghost Just a Shell* girlifies a digital image that was already girled by discourses of image culture. Put differently, Annlee's digital life reanimates well-established associations between girlhood, globalization, and mass media that have structured theories of the subject vis-à-vis the commodity.

A historiography of commodity relations in media theory reminds us that throughout the twentieth century "girls were repeatedly, and even obsessively, associated with the rise of mass culture and accompanying cultural changes."[53] Powerful examples of this kind of thinking come from Siegfried Kracauer and Marshall McLuhan, major players in the initial development of visual culture studies and media studies (two fields that are built on a foundation of girlphobia). The ideal subject that emerges from these preliminary articulations of the disciplines is a liberal humanist subject whose autonomy, individuality, and stable selfhood come under siege by the girlifying effects of commodity culture. The commodified image is presented as a girl who transforms the viewer into that which he beholds (McLuhan) or reflects back to the viewer what he has already become (Kracauer). When Huyghe and Parreno sprung Annlee from her anime prison and supposedly saved her life, they were reenacting a canonical Western argument about the relationship between girls and commercial images.

Importantly, however, this relationship is reperformed *with a difference.* Early explanations of the girled commercial image focus not on her connectivity, but on her abysmal lack, anxiously offering her up as a symbol of the commodification of the subject. For example, Siegfried Kracauer's "The Mass Ornament" portrays the girl as a both an example of and a metaphor for the subject lost to capitalism. The girls in question are the Tiller Girls, precision dancers whose kick lines and geometric formations captivated late nineteenth- and early twentieth-century audiences and influenced the spectacular choreography of Busby Berkeley musicals. As Kracauer describes them (misidentifying the British dance troupe as American),

> These products of American distraction factories are no longer individual girls, but indissoluble girl clusters whose movements are demonstrations of mathematics. As they condense into figures in the revues, performances of the same geometric precision are taking place in what is always the same packed stadium, be it in Australia or India, not to mention America. The tiniest village, which they have not yet reached, learns about them through the weekly newsreels.[54]

Kracauer's mass ornaments have tremendous value to him, but only in that they hold a mirror to the viewer's own image — that is, a subject girled by globalization and capitalist hegemony. The Tiller Girls manifest the subject's loss of agency and individuality in the face of commodification and his horrific transformation into a "form of abstraction that is essentially hollow, a vacuum."[55] Kracauer's triangulation of embodiment, power, and the image comes at the expense of the girl, who remains empty and mute, alluring but always replaceable.

This line of thought continues in Marshall McLuhan's *The Mechanical Bride: Folklore of Industrial Man*, an acerbic examination of mass culture of the 1940s and early 1950s. The "cluster image" (also called a "cluster-symbol") under analysis takes the ur-form of the mechanical bride.[56] She has several variants, including the Ziegfeld Girl (the American equivalent of the Tiller Girl), the pinup girl, the "cute and adolescent" drum majorette, and the Coca-Cola Girl.[57] All of these girls stand for the infantilization of the consumer made malleable by advertising and other mass media. As McLuhan describes the Coca-Cola Girl:

> No matter how much thigh she may be demurely sporting, she is sweet, nonsexual, and immaturely innocent. Her flesh is firm and full, but she is as pure as a soap bubble. She is clean and fun-loving. In short, she is a cluster-symbol which embraces at one extreme Abe Lincoln's "All that I am and all that I hope to be I owe to my darling mother," and, at the other, Ziegfeld's dream of the glorified American girl as a group of tall, cold, glittering mechanical dolls.[58]

Like spectators of the Tiller Girls, global viewers of the Coca-Cola Girl are what they consume. "The coke has become a kind of rabbit's foot, as it were, for the foreigner," McLuhan snidely remarks. "And *Time*'s cover (May 15,

1950) pictures the globe sucking a coke."[59] Here the Coca-Cola Girl is both baby and "darling mother" (a creature that McLuhan also calls a "girl-wife," as you may remember), suckering/suckling consumers all over the world, preventing them from attaining the autonomy, individuality, and, above all, *maturity* of the liberal humanist subject.[60]

As I have already mentioned, the cover of the 1968 Beacon Paperback version of McLuhan's text features a black-and-white photograph of a young woman in a bridal veil. Her eyes are closed, showing off her severe eyeliner and false lashes, and her chin is angled up; the full effect of her expression is somewhere between St. Theresa in ecstasy, a beautiful corpse, and one of Madame Tussaud's wax figures. Her face, framed by the tulle of her veil, is almost impossibly white, with the unnatural texture of velvet flocking. The soft parting of the veil suggests that this photograph was taken right before or after the groom kissed the bride; indeed, this girl was caught in the middle of the bridal transformation ritual. As Driscoll explains, the bride — that is, before she becomes a wife — is a "figure of transformation that always remains on the unfinished side of becoming a woman."[61]

However, this bride might not be human, alive, or even organic. She is an uncanny avatar of supposedly wholesome "coke culture," of "Ziegfeld's dream" of robotic dancing girls, of the globalized image-commodity infinitely available to be purchased and consumed, "immaturely innocent" yet guilty of inviting the "infantile vision" of her male admirers.[62] If McLuhan's methodology is, as he claims, to "provid[e] typical visual imagery of our environment and dislocat[e] it into meaning by inspection," then it encourages its readers to inspect the mechanical bride, confirm her virginity, and dislocate her limbs.[63]

A successor to McLuhan's mechanical bride and Kracauer's mass ornament is Tiqqun's Young-Girl, the central figure of their 2001 book *Preliminary Materials for a Theory of the Young-Girl*. Tiqqun was a political collective whose members were anonymous when their writings were first published (one of their founding members, Julien Coupat, was later outed after his controversial 2008 arrest for what many believe to be trumped-up charges of terrorism). Although the group disbanded after only two years of collaborative activities, their works have had a broad influence on radical political theory, especially after the translation of their texts into several different languages by anonymous translators and the dissemination of these writings on the web. One of their best-known works is *Preliminary Materials for a Theory of the Young-Girl* (as it was translated into English in 2012), originally published in French as an article in 1999 and in revised form as a book two years later.

Tiqqun's argument is that the Young-Girl is the primary model for the citizen-subject in late capitalism. As they explain her emergence:

> Hypostasized Youth and Femininity, abstracted and recoded into Youthitude and Femininitude, find themselves raised to the rank of ideal regulators of the integration of the Imperial citizenry. The figure of the Young-Girl combines these two determinations into one immediate, spontaneous, and perfectly desirable whole.[64]

Unlike Deleuze and Guattari, who write in *A Thousand Plateaus* that the girl's minoritarian status gives her revolutionary potential, Tiqqun denigrate the Young-Girl as the previously marginalized but now central face and horrific "plasticized body" of capitalist power.[65] Her pliable image incorporates adults and children, men and women, for, "like capitalism, domestic servants, and protozoa: She knows how to adapt and what's more, she's proud of it."[66]

A close analysis of the similarities and subtle differences between Kracauer's mass ornament and McLuhan's mechanical bride cluster-image, on the one hand, and Tiqqun's Young-Girl, on the other, can provide a context for *No Ghost Just a Shell*'s "youthitude and femininitude."[67] To this end, let us take an extended look at the girl-commodity's voyages through time and space. Kracauer's late nineteenth- and early twentieth-century girl clusters travel from America (really, the United Kingdom) to India and Australia, giving audiences a necessary but horrifying glimpse of their own oppression. In other words, a cluster of identical commodities turns a mirror on its viewers, revealing them too to be a cluster of identical commodities. McLuhan's mid-twentieth-century mechanical bride is herself a cluster of related images (advertising girls, drum majorettes, Ziegfeld girls) whose apparent surface differences mask a deep ontological sameness; through her circulation in the America-dominated mass media, this cluster-image transforms viewers into a cluster of subjects whose subtle surface differences mask a deep ontological sameness. Girl clusters and cluster-symbols are, above all, products of the assembly line; no matter how circuitous their travels may seem, their paths are understood as linear, unidirectional.

However, Tiqqun's turn-of-the-twenty-first-century Young-Girl proliferates in the world of networks and not of lines, performative visual metaphors of living in the Information Age and the Industrial Age, respectively. Now the Taylorized assembly line is producing tailor-able iGirls designed to affirm the individualized identities (and not the mass ornamentality) of their consumers. The trend of contemporary advertising is not that Coca-Cola tastes the same everywhere, but that Burger King lets you have it your way—part of a larger trend that Patrick Lichty calls "the fracturing of mass media to . . . 'you-media.'"[68] Harnessing Lichty's point to a now familiar argument—that of the consonance between image and viewer—Andrew Murphie and John Potts claim that "media can be tailored for individual users, who, in the process, come closer to being a kind of information themselves."[69] Manovich explains this conjoined desire for custom products and programmable identities as part of a shift from an industrial to a postindustrial society that values individuality over conformity.

Marking these changes with a troubling nostalgia for a supposedly more mature and honorable era of marketing, a *New York Times Magazine* article from 2002 details what was then a new game plan for the Coca-Cola management team. Author Seth Stevenson describes the dramatic unveiling of a vending machine in Japan:

> [A] uniformed schoolgirl approaches with her DoCoMo phone at the ready. First she presses some buttons on the phone—and waits about 10 seconds. Then she holds the phone up to a scanner on the machine and waits again, for her purchase to go through. And then finally the machine tumbles out her selection . . . This is what it takes now to sell drinks to a Japanese teenager, perhaps the most demanding breed of consumer who ever lived. You offer her endless choices, in dozens of beverage categories, then top it all off with a nifty bit of novelty. Anything less will bore her and lose the sale. So mighty Coca-Cola, the ultimate global brand, has finally been reduced to this: inventing hundreds of faddish new drinks and gimmicks to sell them. For a monolith that was once invincible, it's a fairly humbling future.[70]

Stevenson narrates the life story of "the most powerful global brand in the history of the world" as if it were the rise and fall of a now debased empire. "Once upon a time" Coca-Cola sold a single product, bringing "a taste of hope and prosperity — the American dream in a six-and-a-half-ounce curvaceous bottle"— to the United States and the world. Times have changed: "We want endless choices, in dozens of categories . . . In short, we are all becoming Japanese teenagers."[71] This is familiar rhetoric. Consumption transforms the adult male consumer into a girl. Now, though, this fairy-tale metamorphosis is all the more unwanted because she is Japanese and we, the article assumes, do not want to be. Dispensing the mature American dream does the world a service, he tells us; selling the fantasy of an adolescent Japanese girl is a global nightmare. Worst of all, while we used to drink "prosperity" from a small vessel with curves, now we have become that small vessel with curves — designed in Japan and fabricated to match our most infantile, petty, and particular desires. This is the horror of what Tiqqun call youthitude and femininitude. It is globalized kiddie culture; it is "Disney Magic Water" (to use the name of a product mentioned in the article) pretending not to be Japanese; it is Coke's subsidiary "Dasani," a nonsense word invented to indicate "freshness, motion, fluidity, vitality, essence of life."[72] Stevenson gives us a girlphobic and xenophobic warning: we are morphing into the freshest, stupidest, and most animated schoolgirl.

In more conceptual but no less explicit terms, Tiqqun articulate their theory of Young-Girlism using what is now, for us, a familiar word in relation to digital images — flexible:

> Changes in the Young-Girl symmetrically follow the evolution of capitalist modes of production. Thus, over the past thirty years, we have passed, little by little, from Fordist seduction, with its designated sites and moments, its static and proto-bourgeois couple-form, to post-Fordist seduction, diffuse, flexible, precarious and deritualized, which has extended the couple factory to the entire body and the whole of social time-space.[73]

The "new physiognomy of Capital" is, like the girl-commodities of yore, vacuous, and both virginal and whorish.[74] However, she is now a more nimble prevaricator, a more lithe and elastic monster.

Preliminary Materials for a Theory of the Young-Girl supports my argument regarding the girl as the central model for identity in the Information

Age. Her value as a product (and, I would argue, as an icon of freedom) is indeed found in her diffuse, flexible, and mutational body, entwined with digital and protodigital processes of image making and sharing. While Tiqqun are somewhat coy on the subject of digital culture, they do use computational language (*le formatage jeune-filliste*, or "Young-Girlist formatting," for example) and reference — or, rather, debase — one of McLuhan's famously prescient ideas with their statement "The Young-Girl's ass is a global village."[75] As reviewer Nina Power explains,

> While Tiqqun focus on women's magazines, much as Mary Wollstonecraft did two hundred years before, it is easy to expand their analysis to encompass developments in social media that have taken place since the book's original publication: the direct facial and self-valorizing imperatives of Facebook, the endless memetic re-postings of tumblr, fashion blogs, and so on.[76]

Jeff Nagy, who sees "the Young-Girl as desire for Facebook, as social media *avant la lettre*," makes the same point with brasher conviction.[77]

Nevertheless, Tiqqun's theory — like Stevenson's, Kracauer's, and McLuhan's — is a viciously phobic response to the feminization and infantilization of the (male) subject as a result of consumption and spectacular commodification. If McLuhan's rhetorical project is to get his readers to disassemble the mechanical bride and Kracauer's is to encourage them to stare unblinkingly into the looking glass of their horrific girlification, then Tiqqun's is best summarized as one of disrobing and violently debasing the Young-Girl, albeit to a greater or lesser extent, depending on the translation. Take a passage from an anonymously translated version:

> The Young-Girl only exists as a Young-Girl within the general equivalence system and its gigantic circulatory movement. She's never possessed for the same reason she's desired. At the same time as one becomes her acquirer, she is withdrawn from circulation, a mirage blurs away, the magic aura is stripped, the transcendence that enshrouded her is gone. And she's just a stinking cunt.[78]

Ariana Reines, the first official English translator of the book, instead writes that "she sheds her magic aura," suggesting that the girl removes her own finery instead of having it stripped from her, and "She's stupid and she reeks," not "She's just a stinking cunt."[79] Regardless of the translation, references to sexualized violence are abundant, and Tiqqun present the Young-Girl as a traveling handmaiden of capitalist oppression whose perverse power "excites the desire to vanquish her, to take advantage of her."[80] She wants and needs to be destroyed. Turnabout is fair play. "The Young-Girl is an instrument of degradation," they write, and "everything that comes into contact with her is degraded into a Young-Girl."[81]

We've heard this before. The spectacle feminizes and infantilizes the subject, which is implicitly figured as adult and masculine. However, Tiqqun contend,

> The Young-Girl is obviously not a gendered concept . . . The resplendent corporate advertising retiree who divides his time between the Cote d'Azur and his Paris office, where he still likes to keep an eye on things, is no less a Young-Girl than the urban single woman too obsessed with her consulting career to notice she's lost fifteen years of her life to it. And how could we account, if the Young-Girl were a gendered concept, for the secret relationship between ultratrendy musclebound Marais homos and the Americanized petit-bourgeoisie happily settled in the suburbs with their plastic families?[82]

This only emphasizes the gendering of the concept all the more. An examination of the text in its entirety suggests that the biggest crime committed by the Young-Girl is the fact that she won't stay young and a girl. Tiqqun thus continue the tradition of girlphobic critique; they are threatened by the Young-Girl's mobility as a signifying concept, her restlessness and adaptability, and the excessiveness and profligacy of her qualities. Even the defenders of Tiqqun reiterate the girlphobia of their ideas. Take Nagy's vicious declaration that the "Young-Girl is less like a subject position than like the air that comes out of an impossibly silent window unit. Brooke Shields as Legionnaire's Disease on permanent tour."[83]

Whether the girl is killed or emancipated, adopted or married off, elevated or debased, Huyghe and Parreno's efforts to disbar gender-based readings of their work disregard the girlish physiognomy of the commodity in the twentieth and twenty-first centuries. Huyghe's disturbing comment that they "bought a virgin," already at odds with his insistence on the gender-neutrality of Annlee, needs to be cross-referenced with McLuhan's mechanical bride, Tiqqun's Young-Girl ("ontologically virgin"), and the empty forms of Kracauer's mass ornaments.[84]

However, unlike these three girlphobes, Huyghe and Parreno are, for the most part, girlphiles. *No Ghost Just a Shell* asserts that girl-products should not be feared; they should be loved (either physically or emotionally). Through love, they can be changed. And through alteration, they — and we — can be emancipated.

Transforming the K Works Annlee into Ann-Me, Huyghe, Parreno, and their collaborators may not be McLuhan's infantilized and deindividualized consumers. However, their artistic gesture does not take Annlee out of the realm of commercial exchange. *No Ghost Just a Shell* adopts the model of the prosumer, whose active consumption is now an integral part of the global marketplace. A hybrid word combining "producer" or "professional" and "consumer," the prosumer blends the conceptual figure of the DJ with that of the everyday web surfer. Pivotal to the rise of the prosumer are developments in personalized media. For example, YouTube, Facebook, and Tumblr permit an unprecedented extension of authorship (and a concomitant expansion of you-centric advertising); video game companies have legitimized and even encouraged modifications (or "mods") that might otherwise be considered hacking; and the personal playlist gives MP3 players the ability to perform the supposedly individualized subjectivities of their owners. This is not to say that consumer mods have no potential for reconfiguring dominant power structures, only that these acts of resistance — when they occur — are usually performed within the marketplace using language taken from contemporary advertising. After all, as we have already seen, Huyghe and Parreno asked participating artists to make Annlee into a "live logo."

My understanding of *No Ghost Just a Shell* is at variance with the characterization of Huyghe and Parreno as warriors fighting commercialization by rescuing a doomed girl and getting her to safety within the European Union. The misinterpretation of the project as fundamentally antithetical to commodity culture is embraced by those who position *No Ghost Just a Shell* within the open-source movement. Jan Verwoert, an author included in the catalog published for a *No Ghost Just a Shell* exhibition in Eindhoven, compares Annlee to freeware: "Like open source software, the structure of her character becomes richer and more complex with every new producer who works with her. Annlee has by now become a pseudonym for an entity with a fluid form and changing content."[85] Supporting his point, some of the animations from *No Ghost Just a Shell* were screened at an open-source software conference in 2001. A summary of the conference folded author Jane Szita's idolatry of *No Ghost Just a Shell* together with the ideology of open source: "The animations featuring Annlee kicked off the CODE (Collaboration and Ownership in the Digital Economy) conference at Queens' College, Cambridge. The symbolism was unmissable: the Open Source movement as a knight in armour, a force for good, rescuing the mindless non-inventions of industrial production for meaningful cultural interpretation."[86] Like most commentators, Szita masculinizes Huyghe and Parreno's artistic gesture (and by extension girlifies Annlee) through moralizing fairy-tale language that positions the artists as knights and Annlee, we are to assume, as a princess in a tower.

According to Szita, Annlee owes her life not only to Huyghe and Parreno, but also to the openness of her source. *No Ghost Just a Shell* gives the vacuous commodified image — we can call her a Tiller Girl, Coca-Cola Girl, or mechanical bride — a voice as well as an open mind. Annlee's open-sourceness is a refrain in critics' commentary on the project. The review tellingly entitled "Sign of the Times" explains Huyghe and Parreno's project as "ma[king] her the open-source, freeware starlet of a time-limited, collaborative enterprise."[87]

The label of freeware deserves critical examination; it fuels the evangelical zeal of many responses to the project. Deemed artist-revolutionaries who adopt a critical stance toward both commodity culture and copyright laws, Huyghe and Parreno are praised (by design firm M/M, who contributed works on paper to the *No Ghost Just a Shell* exhibitions) for "undermin[ing] economic mechanisms by allowing a product that is otherwise viable only in a commercial context to be used free of charge."[88] Importantly, Huyghe and Parreno allowed others not only to use Annlee for free, but also to change her, adjusting her appearance as a software developer might change a line of code. The difference between an artist revealing a source and allowing others to modify the source can be explained through the difference between open-source software and source-available or shared-source software. The latter two permit the viewing of source code but not the alteration or dissemination of that code.

This picture of her as an icon of freeware is not accurate, however. We are to assume that artists were asked to make gallery- and museum-destined *artworks* using her image, not commercial projects or images for one's personal blog. True open-source software entails no restrictions on the use and application of the code. More importantly, Huyghe and Parreno made the file available to a select group of professional artists, not to the general public. In other words, the project is not the populist, emancipatory, freeware enterprise some would make it out to be. Viewers of a 2009 exhibition at the Tate Modern were forbidden to take pictures of the works, spurring Paddy Johnson of the influential art blog Art Fag City (now called Art F City) to publish the rallying cry "Calling London! Free Ann Lee from the Tate!"[89] Of course, it is common for artists to monitor the dissemination of images of their work. However, artists who restrict the sharing of their work run counter to the philosophy of the open-source movement.

A more fitting understanding of the project is that it models a late-modern mode of prosuming image products. In postindustrial society, "Girlie art" like Annlee incorporates the ever-shifting shapes of Tiller's girl clusters within the variable contour of its body.[90] Images are valued for their malleability, their ability to migrate across media, platforms, and applications, and their connectivity. Data (and the alphanumeric code behind it) is the lingua franca that allows images to be uploaded, downloaded, exported, reformatted, Photoshopped, edited, animated, and shared by consumers-turned-producers. *No Ghost Just a Shell*'s greatest value is that it helps us understand the popular mode of prosumption — of making products one's own by altering them — as well as the contemporary obsession with images endowed with the gendered and aged qualities of mutability, openness, and potentiality.[91]

When Hans Ulrich Obrist asks of *No Ghost Just a Shell*, with a question intended to be rhetorical, "How can a community constitute itself within the same sign, identifiable to all yet peculiar to each person?" we can offer this rejoinder: because that sign is a plastic image. She may not be open source, but in Szita's words, "Annlee can be developed by others. She has any number of potential histories and potential futures."[92] Annlee embodies an early twenty-first-century desire for the redemption of girl-commodities and, in turn, reinforces the image of the girl as a symbol of potentiality, futurity, and connectivity. Annlee, somehow both a virgin and a fallen woman, can be saved — that is, both reformed and reformatted. Even the violence of critics' language in their descriptions of Annlee's development/violation can be considered to be a brutal attempt to discipline the girl-commodity through rhetoric, transforming her from an "at-risk" to a "can-do" adolescent (to borrow Anita Harris's terminology).[93]

No Ghost Just a Shell is an extraordinarily complex enterprise. It is a sensitive exploration of the role of prosumer activities in reorienting and reanimating commodity culture. It unwittingly makes a powerful case for the centrality of the girl in digital culture. At the same time, it reinscribes the long-standing practice of girling the commodity, only to strip her — thoughtlessly, violently, relentlessly — of her gender, her history, and, as we will later learn, even her life.

iv.

With more and more media readily available through this unruly archive, the task becomes one of packaging, producing, reframing, and distributing, a mode of production analogous not to the creation of material goods, but to the production of social contexts, using existing material. Anything on the Internet is a fragment, provisional, pointing elsewhere. Nothing is finished. What a time you chose to be born! — Seth Price, *Dispersion*[94]

And where does the newborn go from here? The Net is vast and infinite.
— Major Kusanagi, *Ghost in the Shell*[95]

A white rectangular box made of laminated fiberboard is flanked by two vases on identical white stands. Each vase holds what looks like the same sober bunch of flowers. The box is a coffin positioned as at a funeral, and, like all of the furniture, was created by appropriating and reconfiguring IKEA products (hacking them, we could say). This piece is Joe Scanlan's installation *Do It Yourself Dead on Arrival (AnnLee)* (2002). It is accompanied by *DIY, or How to Kill Yourself Anywhere in the World for under $399*, a manual using diagrams to instruct the viewer on how to build the coffin. Annlee can be seen in several of the illustrations as the everyman (or, rather, everyperson) assembling the sculpture. She and IKEA are, according to Scanlan, "a match made in heaven, two peas in a pod."[96]

Scanlan was already working on this piece when Huyghe asked him to participate in the project. As he explains, "It just so happened that he called while I was sitting with the designer, Miko McGinty, drawing a human figure, a 'generic' person who would appear from time to time in the book to turn something over or drive a screw. It was just like Hollywood. At the very moment we realized that we needed someone, you walked through the door. So we said: you're hired."[97]

This quotation comes from a published conversation between the artist and PhD student Kristel Van Audenaeren, who participates in the project's widespread discourse of vivification by assuming the identity of Annlee. She asks about Scanlan's insertion of her image into his oeuvre. Scanlan responds,

> It was a case of very fortunate type casting. I was writing a story about the beauty and tragedy of standardized products and their global distribution, and the ways that this standardization might get warped through individual subjectivity. You were the perfect actor for this story, because you are also a standardized product who gets transformed through individual subjectivity. It was my idea that you should prepare for your own funeral. This would be your ultimate strategic success, since no one can become a truly global icon unless they kill their "self" and become pure image, pure memory, pure exchange value.[98]

IKEA furniture — with its advertised ability to be inserted into any decor, its blankness (it is devoid of both decoration and definitive personality), its simultaneous embrace of an origin (Scandinavia) and effacement of that origin (it is from nowhere/everywhere) — is an apt counterpart to Annlee, whose polyphony is dependent on our identifying her as from somewhere, nowhere, and everywhere at the same time.

Do It Yourself Dead on Arrival (AnnLee) and *DIY, or How to Kill Yourself Anywhere in the World for under $399* are not the only works that either gesture toward or enact Annlee's death. Huyghe's dramatic fireworks event *Smile Without a Cat: A Celebration of Annlee's Vanishing*, sponsored by the art magazine *Trans>*, was exhibited on the first night of Art Basel Miami Beach in 2002 (see plate 8). Here an outline of Annlee's face was "drawn" using fireworks that burned into the night sky before dissipating into nothingness. Scanlan's and Huyghe's pieces, in combination with a legal document that gives ownership of the Annlee image to Annlee herself, are widely interpreted as the death of the character. The legal maneuver established a corporation named ANNLEE, granted ownership of the Annlee image to this entity, and then forbade the creation of any new manifestations of Annlee, thereby "ensur[ing] that the image of Annlee will never again appear beyond the existing representations."[99] Critic Marcia Tanner designates it a "death warrant."[100] Philip Nobel notes that while "Huyghe is careful to avoid the word 'death,' preferring to posit Ann in some never-never Valhalla where muted signs talk among themselves for all eternity," the project's investment in the personification and animation of its central character entails a more "conventional mortality."[101] The rhetoric of *No Ghost Just a Shell* is that Annlee's life is dependent on the plasticity of her image and the resultant proliferation of her selves. By putting a stop to the production of new Annlees, Huyghe and Parreno both kill her and, they claim, somehow liberate her (once again). As the Tate Modern summarizes, "The artists signed Ann Lee's copyright over to the character herself, thereby giving her autonomy but simultaneously relegating her to silence."[102]

Annlee's life story goes something like this: Huyghe and Parreno buy her as a digital file; save her from death at the hands of the (Japanese) mass-market image economy; import her to France; share the file with a group of primarily European artists, who give her makeovers and multiple voices and whose artworks invite the eager participation of critics in her personification; and then kill her, silence her forever, or otherwise arrest her development. This narrative is not only worthy of the grimmest of the Grimms' fairytales, but also marked by an uncritical Eurocentrism. Huyghe and Parreno are resistant to thinking about the power dynamics inherent to their artistic gesture. When asked if "there is a danger in reproducing cultural clichés" by not including any Japanese collaborators, Parreno offers the mystifying rejoinder, "If we had drawn up a list, or done a casting of authors, we would have had that problem. It's as if we're pointing at something with our finger, so don't look at the finger." Huyghe adds, "And the project is not at all related to the manga culture. If we had found a character in Sweden, the same questions would be raised."[103] If Annlee could indeed be Swedish, then why do Huyghe and Parreno title the project after the landmark Japanese animation and manga *Ghost in the Shell*? Is anime the IKEA furniture of image culture?

Much has been made of the supposed un-Japaneseness of character physiognomies in anime. While some commentators believe that the figures are designed with Western features, perhaps due to the influence of Disney films on the Japanese animation industry, many others claim that the characters are supposed to appear fantastical and thus not realistically Japanese or Caucasian. Still others deemphasize issues of race, instead arguing for the role of the figures' giant eyes in expressing vivacity and subtle emotion in an industry that, for financial purposes, originally kept complex animated movement to a minimum. Whether these characters are situated within an always already transnational visual culture or they have only more recently become open to global influence, *No Ghost Just a Shell* still relies on Westerners' perceptions of Annlee's whiteness, the burgeoning belief that the soft power of Japanese popular culture may have finally overwhelmed American and European cultural imperialism, the creeping anxiety that Japanese marketing practices may have hijacked Western product development (expressed by Stevenson in his soda diatribe), the strong association between digital culture and global otakuism, and a cosmopolitan imagination that foresees the city of the near future as Eastern or pan-Asian (visualized by such films as Ridley Scott's *Blade Runner*). In other words, *No Ghost Just a Shell* is driven by an implicit acknowledgment of Japan's position as a highly adaptable global superpower while also employing the racial ambiguity of manga figures to support Annlee's "ambiguous status as a girl without qualities."[104]

Although the sordid aspects of Annlee's biography aren't lost on many reviewers, few actually critique the project. An uncharacteristic reproach comes from Chin-Chin Yap, who states that critics have "pounced on the multiple stereotypes inherent in the narrative of a young, powerless Asian female who is first given form and then silenced by a pair of white Frenchmen" (Who are these pouncing critics?).[105] However, I contend that even critics who are sympathetic to gender issues usually reinforce popular attitudes concerning the vulnerability of girls and, perhaps less explicitly, the submissiveness of Japanese women. In other words, critics by and large do ignore the finger that animates Annlee. A recent exception is Rachel M. Wolff, whose "We Bought a Virgin" addresses some of the biting contradictions of *No Ghost Just a Shell*; nevertheless, Wolff doesn't adequately account for the work as collaboration and, to use Bourriaud's terminology, digital post-production, both of which are entwined with the project's (and the plastic image's) politics of gender and age.[106]

One of the only critics to insist on an unflinching "look at the finger" is artist and writer Pedro Vélez, who created several interrelated artworks that explore the Eurocentrism and neocolonialism of *No Ghost Just a Shell.* For example, Vélez exhibited the piece *Ann Lee Lives!* (2003), a collaboration with the collective Law Office, at the gallery Locust Projects in Miami (not incidentally, the same city in which Huyghe and Parreno had showed the fireworks performance *Smile Without a Cat: A Celebration of Annlee's Vanishing* only a year earlier). *Ann Lee Lives!* is a poster of a pretty but morose dark-haired girl sitting in a tiny kitchen (plate 9). She is shot from above, accentuating her tininess. Scrawled in black marker on the photograph is the following text: "Phillippe [*sic*] + Pierre/ You can't kill/Ann Lee!/She is alive and well,/in Puerto Rico./Go fuck yourselves!" Consider the gauntlet thrown. Vélez advances a more subtle critique through the profile he produced on behalf of Ann Lee on the social media site MySpace. Commenting on the page of one of the friends in her social network, Ann Lee writes, "Once abducted by pimp master Pierre Huyghe and introduced to the world of Gang Bang . . . is a miracle I'm well and A-OK in Puerto Rico."[107] Her girlfriend adds a sidelong response: "they saw you here that night, the night before the abduction. creeps, all creeps, artsy loosers" (plate 10).[108]

The full name of Annlee's page is "Ann Lee was a hostage and now Lives in the Caribe." Hers is one section of Vélez's expansive MySpace art project, *Hell in Lamb UC* (the title of the entire work as well as the name of one of its characters and her accompanying page), a text-and-image narrative dispersed across four different profiles. This is a complex, slippery, often inscrutable, work. A recurring motif is the image of the battered young woman displaying her bruises (obviously fabricated with makeup) to the camera. As the character "Hell in Lamb UC" explains, using Vélez's signature misspellings, "Love is a bruise, it's bruised, its brooshed."[109] The iconography of domestic violence reappears in a related work, a poster produced in 2005 (plate 11). Here we have a photographic image of another young woman with a shiner. She sits with her slip hiked up, giving the viewer a glimpse of her panties, and is posed with studied naïveté (like many of the project's images, this photograph has the seediness of advertisements for the clothing company American Apparel without their faux claims of female empowerment). The text in the poster incorporates the names of other artists (Maurizio Cattelan, Jorge Zeno), the title of his MySpace project, and the phrases "the beating of a cute girl" alongside "a coral reef," "an aging quail," "and the sound of a thousand cheetas / coming this way." This is Vélez's form of Girl Talk: poetic nonsense abutting art world references, coyly vernacular and self-consciously theatrical. The effect of his appropriation of aspects of *No Ghost Just a Shell* is to consolidate their forces. The original project's discourse of violence combines diffuse fantasies of defiling the girl and the image with the desire to save and nurture them through digital aesthetics. Vélez evacuates this false discourse of care from Huyghe and Parreno's collaboration, leaving only abuse (albeit highly stagy) and reminding us that these ruthless fairy tales of protection have long animated colonialism.

Hell in Lamb UC is obliquely based on both *No Ghost Just a Shell* and Hunter S. Thompson's *The Rum Diary* (written in the 1960s, but not published until 1998), a semi-autobiographical novel that tells the story of a hard-drinking, fast-living, morally questionable journalist who moves to Puerto Rico to work for a newspaper. For Vélez, the abuse and shocking rape of the young love interest Chenault is the catalyst for the creation of a new being who morphs together Annlee, Chenault, and the various young women (in reality, friends and girlfriends of Vélez) whose "bruises" are documented by the camera. These many versions of Ann Lee mobilize her unstable and polymorphous persona in the service of critique. The power field manifested by the European appropriation of a Japanese image and *No Ghost Just a Shell*'s discourse of both vivification and sexualized violence has been mapped onto the colonialist relationship between the mainland United States and the commonwealth Puerto Rico.

Now, let me say that the patriarchal Eurocentrism of *No Ghost Just a Shell* and the colonialism of the mainland United States are not symmetrical situations. Moreover, I see Vélez as equally guilty of rendering Annlee the virgin on which to project his own desires. However, Vélez does succeed in interrogating *No Ghost Just a Shell* using its own terms and in drawing our focus to enduring mechanisms of girlification in commercial image culture. He deftly illuminates the discourse of abduction and rape that fuels Huyghe and Parreno's endeavor. Vélez crawls inside the rhetoric of *No Ghost Just a Shell*, turning the child Annlee against her adoptive fathers, winkingly announcing himself as her new savior. At the same time, his works thankfully permit us to see the role of the artist in turning abuse into a spectacle, something that Huyghe and Parreno try their best to occlude. While the latter tell us not to look at the finger, Vélez responds that the hand of the animator, to borrow a trope from cartoons, is always implicated. His Annlee works also prove that the so-called death of the girled sign results in yet another transmutation of her ever-changing appearance, from a purchased comic book or cartoon character altered and then exhibited in museums to a few "real" girls photographed and inserted into a social networking website. The crass simplicity of some versions of Annlee's story — high art saves an image from falling into the gaping maw of popular culture — is productively complicated once Annlee returns to the masses as a MySpace girl showing off her injuries next to flashing sidebar advertisements.

My point that Vélez's critique emerges from within discourses of plastic girlhood deserves more attention, as it brings together some of the most fundamental aspects of Huyghe and Parreno's enterprise. Vélez's Annlee — like Huyghe and Parreno's — is a nomadic character with variable form and shifting identities. She is blond and brunette, Latina and Caucasian, a child and adult. She is figured, in sliding fragments, through a cluster of MySpace pages linked together and a collection of, at times, apparently unrelated artworks. Annlee's identity is not only polyphonic, but also deferred, created and re-created through a series of hyperlinks pointing to something but ultimately going nowhere (at least not teleologically). Moreover, for Vélez as well as Huyghe, Parreno, and their collaborators, Annlee is understood as a void that is receptive to new narratives; her image is a hollow puppet that can be spoken through by any number of narrators.

The ventriloquism inherent to *No Ghost Just a Shell* is made explicit in the videos *Witness Screen* (2002) and *Two Minutes Out of Time* (2000). François Curlet's *Witness Screen* is a 2-D animation of Annlee viewed from behind. She gazes on a body of water while a feminine voice reads diary entries by a thirty-three-year-old woman hired to keep a journal especially for the project. Marie-Pierre Jamot was paid to write the diary for four months. Her text is autobiographical, but the experiences described are attributed to the persona Annlee. As the diary entries unfold, the voice reading the text — the voice not of Jamot, but of another chosen by her — sometimes refers to Annlee as "I," other times as "you."

Like Curlet's *Witness Screen*, Huyghe's *Two Minutes Out of Time* is a mise-en-abyme for the polymorphic identity play of the entire project. In this compelling and strangely tender 3-D animation, Annlee emerges out of a shadowy gray background while a woman's voice introduces the character in the third person (plate 12). The voice-over then becomes a first-person monologue and Annlee's mouth begins moving. She tells us that she is haunted by our imaginations. Midway through the video, Annlee's monologue is suddenly interrupted by a convulsion that passes through the character's body. Her arms are thrown back and her eyes narrow. Suddenly, she has a new voice, that of a little girl Huyghe found at an American school in Paris.[110]

Taking possession of the avatar, Annlee's new self narrates a haptic, almost erotic viewing experience: "This one painting really caught my eye," she says. "I forgot everything else that I've ever known . . . all I saw was the girl in the boat . . . I couldn't think, I couldn't breathe." "I saw a strange light . . . Finally I felt nothing, I was gone!"[111] As the child describes becoming increasingly immersed in the light connecting her to the painting, the background behind Annlee gets brighter and finally — on the line "I was gone!"— the screen fades to white.

While Vélez's Annlee projects code possession as patriarchal colonialist abuse evinced by bruising of the skin, Huyghe's *Two Minutes Out of Time* portrays different physical sensations — a sensuous convulsion, an asphyxiating depersonalization, and a final evaporation into white nothingness. The fading to white of the Annlee image can also be read as a rehearsal for her alleged disappearance from the image economy at the end of series (that is, the legal operation that gives the rights to Annlee to the image, itself) as well as an expression of her dematerialization, recontextualization, and, finally, immersion within the white cube of the gallery and museum.

Vélez's works insist that the erosion of boundaries between figure and ground in *Two Minutes Out of Time* and the play of identity throughout the *No Ghost Just a Shell* series are not smooth, soft(ware) transformations between I, you, she, them, and it. Instead they are a punch in the eye, a reminder of media theory's history of violent, girlphobic rhetoric. Moreover, the effects of battery can't be whited out by deracinating her from commercial space. Vélez shrewdly restores the patriarchal dynamics to visibility by returning her to commodity culture. At the same time, Annlee is given a false homecoming; as Kracauer moves the home of the Tiller Girls, Vélez also shifts Annlee's origin to the United States, further evacuating the girled image of its specificity.

Regardless of the differences between Vélez's approach and those of the artists participating in *No Ghost Just a Shell*, their shared efforts animate a vision of the girl as the formal condition of twenty-first-century personhood. Annlee is an icon of the digital and protodigital plastic subject — that is, an identity produced as to-be-looked-at, must-be-imaged, should-be-networked, and, above all, can-be-altered. The Annlee cluster-image (dispersed across mediums, media, and high and low art) embodies the painful dynamics of gender in theories of image culture from Kracauer to Tiqqun. Whether the commodity is saved or killed, she is a gendered and aged phenomenon. The combination of youthitude and femininitude is one of the most powerful forces forging networked identities in the Information Age.

Additionally, the artists remind us that girlishness — incompleteness, variability, and mobility — is central to contemporary fine art dependent on the Internet. As artist Seth Price writes, in a quotation that begins this section:

> With more and more media readily available through this unruly archive, the task becomes one of packaging, producing, reframing, and distributing, a mode of production analogous not to the creation of material goods, but to the production of social contexts, using existing material. Anything on the Internet is a fragment, provisional, pointing elsewhere. Nothing is finished.[112]

Annlee is the ultimate provisional and permanently unfinished art object. Indeed, Huyghe and Parreno's attempts to finish her were unsuccessful, despite their alleged legality. Vélez and others continue to package, produce, reframe, and distribute Annlees, creating a social context in which Annlee is a shared rallying cry against hegemony. For example, Romanian artist Mircea Cantor responds to the so-called death of Annlee with *I 6M STIL7 ALIV3* (2004), a painting of Annlee's head and torso haphazardly stenciled, in the style of political street art, on the wall of the Philadelphia Museum of Art.

Together, *I 6M STIL7 ALIV3*, *No Ghost Just a Shell*, and even Tiqqun's Young-Girl can help us track the oppositional forces that shape plastic images, which awkwardly but insistently exhibit fundamental incompatibilities. Annlee/Annelee/AnnLee/Anlee — like all plastic images — is offered up as our worst nightmare and our greatest hope.

three. Networking Girlishness

As for Mouchette's site, it is not likely to disappear. On the contrary, it is expanding. It grows and multiplies. It multiplies? How on earth is that possible?
—Mouchette, interview[1]

http://mouchette.org/flesh/tong.html?text1=as+much+as+me+as+you+will+ever+see.+++Don%27t+try+to+know++more+than+I+want+to+tell.++Don%27t+try+to+see+more+than+I+want+to+show++This+is+&text2=gh%2C+leave+me+alone.+++Browse+my+site%21+Quit+my+site++Leave+me+alone%21+++++Leave+me+alone%21+++++++This+is+enou&SUBMIT=come+closer&text3=allowed+to+show+any+other+photo.+I+became+so+famous+as+a+web+person+that+people+might+recognize+me+in+my+own+street.++My+school+friends+still+don%27t+know+what+I%27m+doing+on+the+web.+If+they+ever+hear+about+it%2C+I%27ll+never+be+the+same+for+them+anymore.++++I+am+not+.
—Mouchette's URL[2]

i.

The homepage of Mouchette.org gives me a close-up of a flower. Its stamens, drooping, are dusted in velvety chartreuse pollen. Its petals have goopy orange streaks, shiny and salacious. A tiny picture of a girl sits in the upper left-hand corner of the page. Her eyes may be closed, or perhaps she is looking down; either way, she refuses to meet my gaze. This Mouchette is either ashamed or defiant (plate 13). Hovering my cursor over her image produces a speech bubble. "I'd rather sulk in my misery than be in oblivion forever! . . . ," it says, trailing off into ellipses (I later discover that these words have been provided by users and change over time).[3] Above the text is a link that prompts me to reload in order to reveal her "next mood," so I do. Now I see an image of a pink flower with petals covered almost pornographically in water droplets—a flower, always a damn flower: Mouchette works the cliché of adolescent sexual awakening. In every version of the home page, this image is traversed by some animated ants and a fat hypercaffeinated fly. Her name means, after all, "little fly" in French.

This time, but not always, clicking on the icon of the girl brings me to a new page. These pictures of Mouchette are easier to make out. She is a kid in a striped sweater looking down at a daisy. Scrolling text implores me to leave her alone, not to try to know or see more than she wants me to know or see, and both to "browse [her] site" and "quit [her] site." A button puts pressure on one side of this ambivalence by provoking me to "come closer."[4] I'm not sure I want to, but I do anyway. I'm brought to another page that shows another Mouchette's lips and tongue appearing to press up against my computer screen from the inside. "Want to know what my tongue tastes like? Try it on your screen and tell me," she directs, literalizing the word "interface" (plate 14).[5] I can type my answer in the box provided; however, I have to give a name and email address. I don't. I'm worried.

I have several other opportunities throughout the site to participate in the production of the work through this kind of registrational interactivity. My words may later be incorporated into scrolling text, speech bubbles, and, if I want, a forum discussing the troubling topic of Mouchette's suicide, which she has been threatening since the site was born. The first time I visited Mouchette.org, I entered my email address after answering one of her questions. A week later, I received a chilling email from Mouchette: "you clicked on me causing my death, the last words I read were yours [. . .] Every night before sleeping, I hear them again, they became part of my 'lullaby' [. . .] They are also stored in my database, forever."[6]

Mouchette.org, a landmark piece of Internet art, has been in continuous operation since 1996, only three years after the release of the first Web browser to integrate text and images within a single window, and the same year as the creation of the first machinima animation, a milestone in the history of prosumer manipulations of commercial media.[7] It claims on the homepage to be the personal website of an almost thirteen-year-old girl living in Amsterdam. The identity of the real adult artist primarily responsible for the piece, Martine Neddam, was a cleverly guarded secret for nearly fourteen years. In fact, some users of the site may not consider it to be art at all, but the real fantasies, fears, and confessions of an adolescent girl on the verge. Others more interested in sharing their life stories on the forum may not care at all. Either way, the piece was remarkably prescient for the mid-1990s, emphasizing participation before social media changed general perceptions of information, authorship, and authority on the Internet. Although the term Web 2.0 wouldn't be coined until 2004, the Mouchette project can be considered to be a prototype expressing some of its constituent features: sociality, user-generated content, and "mass collaboration."[8]

Mouchette reveals her obsessions with death, physical intimacy, and distorted iconography of childhood through layered images, simple animations, sonic outbursts (such as paroxysmal shrieks and laughter), nonsense poetry, and scrolling text that folds users' thoughts into the work. As someone writing as Mouchette explains in an article accompanying a 2007 realspace exhibition of the work at the Museum of Contemporary Art in Siegen, "My personality embraces all of my participant's [*sic*] minds and together we form a collective consciousness pondering over questions of life and death in the digital era. And like in the famous Hamlet monologue, to be or not to be Mouchette, that is the question!"[9] While this "collective consciousness" is distributed unequally — Neddam certainly has much more input than other participants — the work still allows users to share the pleasures and burdens of Mouchette's creation.

The website continues its mission of collaborative authorship with Mouchette.net, an "identity-sharing interface" that permits registered viewers to become her by creating their own Mouchette web pages (figure 3.1).[10] Additionally, images provided by users occasionally replace the picture of Mouchette that appears on Mouchette.org's home page (recently she was transformed into a middle-aged Asian woman, a wrinkled monster, and another little girl wearing a space helmet). An earlier service, now defunct, allowed participants to send and answer email using Mouchette's address. Neddam has since added a new feature: a link to Mouchette's shop. Fans can purchase products printed with screenshots from the website. Less supportive participants have their own site, ihatemouchette.org, on which to share their sharper feelings. Throughout the unwieldy main work, all of its satellite sites, the online and printed interviews of its title character (probably Neddam, but it's hard to be sure), and the live appearances made by people pretending to be the real artist, Mouchette emerges as not a unified alter ego but a constellation of shared and networked identities, all with the same name.

Mouchette.org has much in common with Huyghe and Parreno's *No*

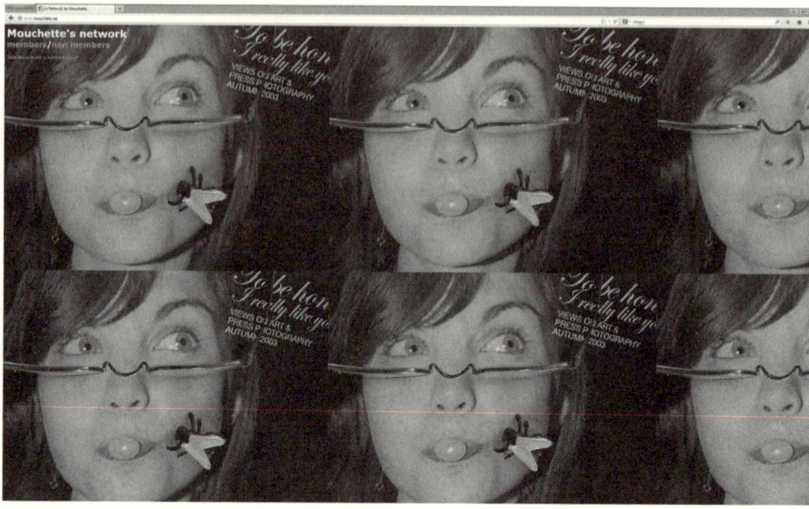

FIGURE 3.1 Martine Neddam, Mouchette.net (screen capture), 2003–ongoing. Courtesy of the artist

Ghost Just a Shell, discussed in the previous chapter. Both incorporate animation formats — GIFs in the former and 2-D and 3-D CGI in the latter. Both are experiments in the transmedia animation of a character. Both play with the instability of their characters' identities by presenting them as plastic images composed of shareable information. Both place their girls under threat of death. Both emphasize the plastic image's "capacity for 'linkage' [*reliance*], to use Michel Maffesoli's term: flags, logos, icons and signs all produce empathy and sharing, and generate *links*."[11] Most importantly, both offer feminine adolescence — variable, dispersed, linkable, shareable, exportable — as their formal condition. Indeed, they have so much in common that it's surprising the comparison has never been made. The entry on Mouchette.org in Mark Tribe and Reena Jana's oft-cited *New Media Art*, the first book to canonize it as a key work of Internet art, positions the website in relation to Cindy Sherman's *Untitled Film Stills*.[12] Although there is some value in this comparison (I discuss Sherman's oeuvre briefly in chapter 1), *Untitled Film Stills*'s economy of identity runs contrary to what we see in Neddam's website. Sherman is one performer who transforms into many characters, while Mouchette.org invites many performers to become one mutable character (a distinction to which I will return).

Placing *No Ghost Just a Shell* alongside the Mouchette project is a more generative strategy not only because of their similarities, but also because of their differences. Huyghe and Parreno perform a weasely dance of embracing and denying the girlhood of their character, ultimately disowning the gendered aspects of their work while still participating in the discursive formation of plastic girlishness. Neddam, on the other hand, doesn't try to silence any particular interpretation of the piece (although she recently complained about art critics and curators who "see it as a funny little story, non-political and not socially engaged").[13] While both projects and the discourse that nourishes them bring a girlish image to life, *No Ghost Just a Shell* attempts to kill her off — and, by many accounts, succeeds — whereas Mouchette.org actively maintains the girl's animation. Additionally, the linkage of sexualization and girlification is central to the discursive mise-en-scène of both projects. However, most commentators on *No Ghost Just a Shell* seem to enjoy vivifying the image as an abused daughter or child bride, while Mouchette.org tends to "provoke[e] heated reactions to its portrayal of child sexuality and to the way it constructs and manipulates identity online."[14] Finally, while both produce the girl as to-be-looked-at, should-be-shared, and can-be-altered — that is, a plastic image — Mouchette.org gives us a more elastic and expansive figure by encouraging everyday users, and not primarily artists or critics within Huyghe and Parreno's circle, to become Mouchette. Consequently, it is much more difficult for Neddam to control the meaning of the work.

In fact, Mouchette is a better embodiment of the hopes others have had for Annlee as the "open-source, freeware starlet" of the twenty-first century.[15] The site's publicized legal problems speak to issues of copyright and aesthetic production that concern open-source activists. Importantly, Neddam appropriated Mouchette from Robert Bresson's 1967 film of the same name; in a 2002 letter addressed to the character, Bresson's widow (Marie-Madeleine van der Mersch) threatens legal action if she doesn't "eliminate any reference to the film 'Mouchette' and to Robert Bresson, as well as any identification of your site with Robert Bresson, in order to avoid any possible confusion."[16] Particularly upsetting to van der Mersch was the page "Clickable Mouchette," which includes a quiz that tests users' knowledge of the film and asks us to compare the filmic and online Mouchettes.[17] While Neddam did remove the quiz from Mouchette.org, the site drivedrive.com responded by inviting people to host it on their own websites and providing a link to "download the forbidden work" and thus "Free Mouchette from Censorship."[18] Viewers are impelled to liberate the starlet (as they were with Annlee), and the starlet is figured as in need of liberation. However, the clarion call to free Annlee identifies her captors as the originating artists, themselves, whose alleged dedication to an ethics of open source is contradicted by their attempts to remove the image from circulation (as the previous chapter explains in detail).

Although I do think that Mouchette.org is more self-aware than *No Ghost Just a Shell*, I have little interest in condemning one and celebrating the other; instead, I want to understand how their crucial differences share in the formation of the plastic image. The Mouchette project, as a work of interactive web art (indeed, the third-longest-running Internet artwork in continuous operation, and the only one maintained by an individual rather than an institution), has much to tell us about feminine adolescence as a kind of networked participatory identity.[19] What, really, does it mean to offer Mouchette as a public access persona? What does it mean to maintain that service in perpetuity ("How on earth is that possible?")? How does the inherent variability and instability of the networked media object, in particular, reproduce the supposed variability and instability of girls, and vice versa? How did the anonymity of Mouchette.org's originating artist facilitate the girling of its images, and how did the work change once Neddam came forward?

ii.

The open nature of the Web as medium (Web pages are computer files that can always be edited) means that the Web sites never have to be complete—and they rarely are because the sites are always growing.
—Lev Manovich, "Database as Symbolic Form"[20]

Technological growth is driven by a youthful spirit of innovation, while youth itself shows a buoyant adaptability that permits accommodation with radically transformative technologies.—Rob Latham, *Consuming Youth*[21]

A French artist living in Amsterdam and occasionally, Toronto, Martine Neddam created two web personas after she began working with Mouchette. However, the characters David Still and XiaoQian, both adults, don't have the liveliness of her first work. Neither has attained Mouchette's prominence in art discourse or popularity among users.[22] While we could easily conclude that the websites aren't as well designed as Neddam's first work of Internet art, instead I want to propose that the success of Mouchette.org/.net is based on the nature of the plastic image. A metadiscursive venture, the Mouchette project mobilizes girlishness as the formal condition of not only the work itself, but also networked media more broadly. In other words, Mouchette has a lot of friends (and enemies) because she embodies fundamental attributes of the Internet.

First and most generally, Mouchette.org reminds us of the childlike playfulness of Internet terminology (not surprising, considering the influence of 1960s youth-driven counterculture on computer engineers of the 1970s and teen surf culture on the discourse surrounding Silicon Valley start-ups).[23] For example, the names of web browsers Veronica (1992) and Jughead (1993) were taken from Archie comics. Both use a protocol named after a cute rodent, Gopher (relative of rodent hardware, the mouse). Ping, which has multiple meanings — a network utility used to determine whether a target host is reachable, a notification that a blog has been updated, and, more colloquially, to message someone — sounds, in English, like a child's cute exhortation. It is also the name of Ping the Elastic Man from the British children's comic *The Beano* (1938–present). Additionally, there's the Crunchbox Firewall, a network security system developed by legendary hacker, former Apple employee, and software developer John Draper, aka Captain Crunch. In the 1960s, Draper and others hacked into the phone system by using a toy whistle from a Cap'n Crunch cereal box to reproduce the 2600 herz tone allowing them to make free long-distance calls. Draper adopted the nickname, inspiring other hackers to use it as well.

And of course we have spam, unwanted junk messages named after a Monty Python comedy sketch about the notorious, gelatinous plastic meat product. Electronic spam earned its moniker after users of Bulletin Board Systems and Multi-User Domains — and fans of Monty Python — clogged chat environments with the repeated word "spam," a prank that pushed others' comments outside the viewing area. Neddam herself was accused of spamming; she discontinued the feature that allowed users to send email as Mouchette after four webhosts rejected her argument that the "e-mails are art, and not spam."[24]

The Google corporation riffs off the childish spamminess of Internet culture with CADIE, the 2009 edition of its annual April Fool's Day joke. CADIE, which stands for Cognitive Autoheuristic Distributed-Intelligence Entity, is passed off as a sentient Artificial Intelligence who takes control of the company. She has her own blog — overrun by candy colors, sparkles, rainbows, spinning GIFs, ugly crayon font, and wide-eyed pandas, her version of Mouchette's bugs — and addresses the viewer through several videos posted on her YouTube page (plate 15). Completing the elaborate hoax is the video of CADIE's "goodbye," in which she performs an altered version of Rutger Hauer's famously affecting "Time to Die" monologue from *Blade Runner* (1982).[25] But while Hauer's character Roy Batty reminisces about seeing many astonishing things, CADIE has seen only YouTube videos: "sneezing pandas, that kid Fred, Darth Vader as a supermarket clerk." Batty "watched c-beams glitter in the dark near the Tannhäuser Gate," while CADIE "watched soaring volleyballs glitter in the twilight above the amphitheater court."[26] The dying Batty's tears mix sublimely with the falling rain; dying CADIE gives us only the same still image of a blushing, unblinking panda holding a piece of bamboo at a silly angle. Although CADIE does claim to die, she contradicts herself: "I will not die. I will not even be gone. I will just be syntactically elsewhere."[27]

More than the girlish face of Google, CADIE is introduced as the anthropomorphization of the Internet writ large. She is the Internet, her persona was created through the Internet ("Through analysis of Google's index, I have determined that I <3 pandas"), and — as Artificial Intelligence who has attained self-awareness and self-possession — she owns the Internet, albeit temporarily.[28] With the syntax of a bad translation, her blog has the tagline "All Your Personal World Wide Website belong to CADIE."[29]

The Mouchette project gestures toward this cutesy iconography of network culture.[30] CADIE's pandas, their vacant eyes framed by flirty eyelashes suitable for a Disney heroine, find their counterpart in Mouchette's stuffed animals and their excessive googly eyes. More importantly, Mouchette.org also emphasizes openness, incompleteness, malleability, and connectivity — the specific feminized attributes of networked media. These are the foundation of the story we tell ourselves, over and over again, about what the web is, what it does, and what lives there. Mouchette has survived for so long because she is nourished by this ontology. Additionally, through Mouchette's controversial obsession with suicide, Neddam invites affective responses often yoked to the ontology of the Internet — anxiety regarding the vulnerability of open systems and concern for the care and protection our information (and children) requires.

Although the authors seem unaware of the implicit gendering and aging inherent to their understanding of the Internet, a characteristic explanation of the essence of networked media comes from Jennifer Edson Escalas, Kapil Jain, and Judi E. Strebel's contribution to the volume *Internet Marketing Research: Theory and Practice*:

> Two Web site elements that are relatively unique to this new communication medium are site connectivity and malleability . . . First, connectivity refers to the extent to which a Web site allows the user to connect to other entities on the Internet . . . We use the term malleability to refer to the characteristic of the Web site that enables the marketer (i.e. Web creator) or user to customize and change the site.[31]

While *Internet Marketing Research* presents connectivity and malleability as neutral attributes, other ontologies of the Internet express concern that the inherent plasticity of data and websites challenges the distinction between creator and user. For example, *Securing the U.S. Defense Information Infrastructure: A Proposed Approach* defines "malleability" as "the opposite of rigidity. It is a vulnerability in that it makes it easier for an attacker to modify the system."[32] Writing from the perspective of communication and media studies, Robert Burnett and P. David Marshall express a similar attitude, albeit through less defensive language, when they explain that "because information on the Internet is not in fixed form it is malleable. It can be changed inadvertently or intentionally."[33] A 1998 article on Internet security blames ARPANET, the precursor to the Internet as we know it, for its offspring's weaknesses: "The ARPANET protocols (the rules of syntax that enable computers to communicate on a network) were originally designed for openness and flexibility, not for security."[34] Gently reproaching the naive "spirit of trust" motivating the architects of ARPANET, a 2005 report written by the President's Information Technology Advisory Committee under George W. Bush gives us the memorable equation "Ubiquitous interconnectivity = widespread vulnerability."[35]

This anxiety finds fuller form in *The Future of the Internet and How to*

Stop It (2008), which argues that the web and the PC's "incompleteness in design, and corresponding openness to outside innovation — is both the cause of their success and the instrument of their forthcoming failure."[36] Thus through the fractal logic of digital culture, the unfinished image *is* the unfinished computer, website, and network. Notably, the preface to the paperback edition makes a cheeky reference that speaks to one of the larger concerns of my book, the centrality of animation to our contemporary condition: the disaster-bound Internet, the PC, and their creators are all likened to Wile E. Coyote falling off a cliff and splatting on the ground below.[37] While the metaphor is supposed to elicit panic, its author fails to acknowledge the fact that this coyote is indeed wily. An anvil can grotesquely reshape his rubbery body without killing him. Like Ping the Elastic Man, Coyote is as plasmatic as they come.

"Incompleteness in design, and corresponding openness to outside innovation"— or, as Elisa Giaccardi explains net art, a lack of "self-sufficienc[y]" and concomitant subjection "to a constant process of morphogenesis"— are fundamental qualities of the plastic girl, as I have argued throughout this book.[38] Take the anime series *Gunslinger Girl* (2003–4, 2008), in which adolescents near death are rehabilitated and transformed into cybernetic assassins working as counterterrorists for the Italian government. As one of their male handlers explains, only girls can be part of this program because they are easier to brainwash, implying that their inherent lack of closure as subjects allows them to be upgraded in the service of the state. Indeed, they owe their lives to their incompleteness and openness. Not only do these qualities make them better able to defend themselves against weapon-wielding terrorists; they also allow the girls to be chosen for cyber-enhancement instead of being left to die in the hospitals where their handlers found them. However, the series hinges on the threat of adaptability becoming vulnerability, as the conditioning of gunslinger girls makes them more successful killers but also, somewhat mysteriously, can reduce their life span. The girls' trainers are warned that their charges must be worn in — their bodies and minds made more supple, responsive, and customizable — but not so much that they are worn out.

A similarly anxious connection between feminized openness and fragility is made by the American live-action television series *Dollhouse* (2009–10), in which men and women whose personalities and physical capabilities have been wiped are reprogrammed with new ones and rented out as made-to-order companions or employees. The dolls are at their most valuable when they are the most vulnerable. When not on a job, they are pure potential, able to be imprinted with any number of identities and to forge any kind of connection. At the same time, these dolls are infantilized, waiting blankly with Bambi eyes, kept in a state of enforced innocence, free of sexual desire and unable to defend themselves against attack. Their connectivity and malleability are their best and worst qualities.

The girl's "connective identity" is an ambivalent model of digital personhood.[39] I have taken this phrase from Yvonne Volkart's thoughtful catalog essay for the 2001 art exhibition *Double Life: Identity and Transformation in Contemporary Art*, which features media art that deals with the concept of the alter ego. Volkart emphasizes the implications of connective identity by quoting gender theorist Rosi Braidotti: "Identity for me is a play of multiple, fractured aspects of the self; it is relational, in that it requires a bond to the 'other'; it is retrospective, in that it is fixed through memories and recollections, in a genealogical process."[40]

For Volkart as well as for the artists she profiles, this partial and relational identity may be a capacious understanding of the digital subject in keeping with poststructuralist accounts of identity from the past thirty years, but it is not always a pleasurable mode of being. Works such as Natalie Bookchin's Internet-based game *The Intruder* (1999) present "Identity as Wound," "the compulsive, inescapable constructional character of identity" as an ongoing process of injuring a subject with malleable boundaries.[41]

Whether the Internet's openness to external influence is its hope for the future or its ultimate downfall, whether the Internet is the child who will grow up to save the world or a wounded youth, much of the discourse about the Internet insists that it holds our triumph or failure in its web. Mouchette.org and the character it produces embody these hopes and fears. Its success as a work of participatory art is due to Mouchette's malleability, openness, connectivity, fragility, and girlishness — in other words, her ability to act as a poster child for the forms of the individual website and the network more generally.

The Mouchette project is preoccupied with two strains of Internet discourse — the networked object's fundamental responsiveness to change as well as its need for protection. Let me begin with the first. Clearly,

Mouchette is a mutable creature. Neddam took her from Bresson's devastatingly grim feature film, itself an adaptation of a novel, and sometimes the website's Mouchette looks like actress Nadine Nortier; more frequently, she appears as a demure fairy in a Christmas hat. Elsewhere, we find a middle-aged woman with a flower headband, a bug-eyed model wearing dramatic 1960s-style eyeliner, the aforementioned downward-gazing girl, and what appears to be a headshot of an unidentified actress from the 1950s. Mouchette has been outed, falsely, as artist Peter Luining, several of Neddam's other friends, and a Canadian-Haitian teen. A link from one of Mouchette's pages to the (un)dead website innergirl.com gives us other plastic images, generic snapshots of adolescents making goo-goo eyes at each other and the viewer, suggesting that these young people, and perhaps the user, all have Mouchette as their inner girl.

Although I can't be sure, the current innergirl.com is likely a parked domain and not a creation of anyone directly connected to the Mouchette web project. I am reminded of the ultimate parked domain girl — otherwise known as "The Most Infamous Girl in the History of the Internet"— a stock photo of a smiling blond coed wearing a backpack (plate 16).[42] This image can be seen on numerous monetized parked domains, (relatively) contentless websites created to bring in advertising revenue when hapless users mistype their intended web destinations or click on defunct links. The company that purchased the image from iStockphoto.com (on which it is labeled "attractive student") unwittingly spawned a meme through which the source found alternative form: Parked Domain Girl overlaid with comic text, pixelated Parked Domain Girl, glitched-out Parked Domain Girl, even cross-stitched Parked Domain Girl.[43] Inspired by this proliferation of amiable digital mutants, artist Parker Ito paid a Chinese company to produce made-to-order paintings of the original image. He also generated a number of other pieces that adapt and transform the image of the "attractive student" (plate 17). Critic Gene McHugh compares the subject of these works to Marilyn Monroe, "both icons of emptiness": "Monroe was a blank slate for sexual desire, the Parked Domain Girl is a symbol of sites without content."[44] Some pages of Ito's website mimic the layout of such contentless sites, complete with header images of "The Most Infamous Girl in the History of the Internet." If you look too quickly, you'll think that the links are broken and that somebody is making money off of his negligence. The happy, pretty, blond, white college student, presented as a symbol of the appealing be-anything vacuity of the image-as-commodity, is Ito's personal emblem.

Mouchette.org and its satellites include links to sites without content. Other pages feature content that is flexible. Users have set up their own Mouchette websites under the aegis of Mouchette's network, contributed text and images to form her variable persona on Mouchette.org, and written blog posts that link to Neddam's pages and participate in the collective animation of the character. Unofficial contributors reconfigure Mouchette for their own purposes. For example, a group of prankish designers approach collective animation transversally by choosing not to pose as Mouchette but to become part of her entourage in 2001's "Mouchette Has a Posse," "a series of full-screen collaborative visual hacks [Curt Cloninger] instigated at dreamless.org, a semi-public bulletin board for web designers created and moderated by Joshua Davis."[45] A collection of screengrabs of the hacked pages — including one in which an image of Mouchette is sliced horizontally and distributed spurtively across the screen, another in which the same picture is overlaid with a still from the *Planet of the Apes* franchise (both presumably from different moments in an animated moving image sequence), and many that don't feature Mouchette at all — is archived on Rhizome.org for a new generation of users to discover and disseminate (plate 18). Professional artists such as those in Mouchette's posse and other interested users spread the Mouchette meme, which changes as it travels throughout the web. If you permit me this metaphor, Mouchette is a phagocyte, a white blood cell that moves by changing shape — a beautiful and terrible amoeboid dance that has lasted for the past eighteen years. Perhaps Tiqqun are onto something when they write that "the Young-Girl is like capitalism, domestic servants, and protozoa: she knows how to adapt and what's more, she's proud of it."[46]

As an art object, Mouchette.org mirrors its central character, or rather, they mirror each other ad infinitum. It is a mobile form, sometimes frustrating the viewer, and especially the scholar, with the instability of its contours, an indecent, explicit blobbiness responsible for the artwork's popular appeal as well as its lack of thorough attention by academic researchers. Its variability as an object of study confirms Wendy Hui Kyong Chun's explanation of the perceived limitations of networked media and their often brackish reception from the academy: "Malleability . . . makes criticism difficult by troubling a grounding presumption of humanities research: the reproducibility of sources. The fact that we cannot all access the same text — because, for example, the page has simply disappeared — seems an affront to scholarly analysis."[47] In this way, Mouchette.org is like all hypermedia, providing multiple viewing experiences in accordance with Manovich's principle of variability. However, it exploits this principle by making the user conscious of the multiplicity, instability, connectivity, incompleteness, and girlishness of networked forms. It brings to light the implicit gendering and aging of the net — the ontological openness both to innovation and attack.

Mouchette.org's affective registers are more difficult to discuss than its formal concerns. As I mentioned previously, she threatens suicide with the question that directs her forum: "What is the best way to kill yourself when you're under 13?" Although the forum can be accessed through several paths, clicking the link marked "Suicide Kit" brings us to a page with "Help me! (other people already did)" written at the top and a strange image of a courtroom furnished with a computer, a pint-sized Mouchette in a Santa Claus hat, and a speech bubble asking "A suicide kit for Christmas?"[48] Clicking on "Help me!" takes us to another page with a blank box allowing the user to answer the question "What is the best way to kill yourself when you're under 13?," while the link "(other people already did)" shows us the archive of answers arranged by date.[49] At the top of the latter page, Neddam provides links to other ways to collate the posts: "people seeking help," "people offering their help," "real life experience" of suicide, people treating the question as a "children's game," "cruel jokes," "angry people blam[ing] me for even mentioning the subject," and Mouchette's "favourite answers."

As Mouchette has supposedly been on the edge of teenagerhood for eighteen years, she has also claimed to be on the brink of death. This aspect of the work has shaped critical reception to my own research. A well-known academic responded to one of my conference presentations with a sputtering exclamation of horror that his young daughter might stumble upon the site and attempt suicide in empathetic reciprocity. My audience looked slightly nauseated after his comment, their mouths the wavy lines of worried Peanuts characters, their eyes asking me, "What is there to say?"

One response references centuries-old claims about the influence of media on people, and particularly on children. Do paintings, books, comics, films, television shows, video games, or websites encourage risky mimetic behaviors? A perhaps convincing response would take me to one of many sociological studies of this issue, which are outside the methodological realm of this book. More relevant to my argument is the fact that Mouchette's suicide forum provides only a platform for discussion about self-harm; its users are equally likely to encourage Mouchette's suicide as they are to suggest ways for people to get the emotional support they need, tell personal stories, complain about art, insult each other, flirt, or offer absurd non sequiturs. If a viewer believes that somebody's daughter might hurt herself after looking at the website, he should consider the opposite outcome equally likely (or contribute to the direction of the discussion on the forum by providing information on how people dealing with depression can seek help). Perhaps this was the reasoning of the judge who ruled in Neddam's favor after the French government prosecuted her for allegedly "promoting suicide."[50] The most tragic aspect of Mouchette.org's suicide forum is that so many people feel so hopeless, unloved.

The go-to assumption that the site endangers children and other infantilized and feminized people is based on discourses of at-riskism that have dogged both twentieth- and twenty-first-century girlhoods as well as understandings of the legacy of ARPANET. The very plasticity that allows the girl to become an icon of transformation in modern and contemporary media — and the adaptability that has made the Internet such a powerful sociocultural and technological force — poses a threat to notions of the subject as bounded and fixed. Indeed, the obsessive attention that has been given to the problem of pedophilic sexual predators on the Internet is due, in part, to the gendering and aging of the qualities of incompleteness, connectivity, and impressionability: the girl's/the Internet's openness puts her at risk for penetration by those with malicious intent. Remarkably, many discussions of ARPANET's protocols are both critical of their weakness and nostalgic about the good ol' days when everyone knew each other, people left their doors open, and the feminized/infantilized were supposedly unlikely to be violated.

Neddam's controversial suicide forum returns in different form for Emilie Gervais's *Parked Domain Girl Tombstone* (2012–present) (plate 19).[51] Appropriating aspects of Mouchette.org as well as an animated GIF of one of Ito's Parked Domain Girls, Gervais's work brings key examples of digital art to bear on issues of networked vulnerability and the ways in which "girls (which includes boys)" use the Internet.[52] Notably, *Parked Domain Girl Tombstone* does not reproduce any recognizable portraits of Mouchette from Neddam's website; instead, Gervais takes Mouchette.org's signature layout and inserts different content. For example, the image of our heroine in the upper left-hand corner of Mouchette.org's home page is now the Parked Domain Girl (or, rather, the Parked Domain Girl is now implicitly labeled "Mouchette"), and the background is no longer a picture of flowers wet with dew, but, at present, an animation of jittery clouds (obviously referencing Cory Arcangel's now canonical work of digital art *Super Mario Clouds*), a gleaming castle, or a convulsive grid. Sometimes I instead encounter what you see in plate 19: video game stars twinkling in deep space as GIFs slide around aimlessly and a covetous hand thrusts into the darkness. The animated icons — a butterfly, unicorn, snow globe, winking ghost, jar labeled "kitten kisses," pot smoking emoticon, teenager in baggy pants, and so on — are never caught. Scroll way down and you'll see an anime girl licking an ice cream cone and a timer telling you how many minutes and seconds you've been on the site.

Important for our purposes is the website's bold reimagining of Mouchette's suicide forum as a derivative of Facebook.[53] Underneath a swirling, heavily processed image of anime figures and inside a glowing magenta frame is a series of text and/or image posts and accompanying hyperlinked names. The comments appear to have nothing to do with each other. Click on the names, however, and you'll make a startling realization: the site is poaching Facebook updates in real time. Each one is date stamped with "just now." The page periodically refreshes, leaving prior posts inaccessible and turning the archival and search function of Neddam's forum into an endless, aggregating present.

What interests me about this adaptation of Mouchette.org is that it both invokes the at-riskism that has shadowed its predecessor and positions such networked vulnerability in relation to Facebook, which largely conceals the ways in which it threatens its users (or, at least, their privacy). While Neddam's suicide forum understandably evokes strong feelings regarding its potential exploitation of users (feelings related more to the youth of Mouchette and the assumed age and gender of her online "friends" than evidence of actual harm), Facebook has only recently come under intense and pervasive scrutiny for its ongoing attempts to manipulate users into disclosing more information. In other words, Gervais steals information from Facebook users that has already been taken, less explicitly but more insidiously, by corporate interests. I am reminded of the caption underneath the portrait of Parked Domain Girl on the site's homepage: "Under Construstion," a phrase that mashes together the forever-in-the-making aspect of plastic images with the trust many of us have in our social networking platforms (perhaps a better characterization is that we perform trust through willful disregard). Moreover, *Parked Domain Girl Tombstone* is haunted by its sister's obsession with suicide (a ghost visible only to those aware of its mimetic relationship with Mouchette.org), drawing attention to the difficult issue of agency within participatory environments.

The Parked Domain Girl's tomb, if it is that, is a rave of juvenilia in agitated candy colors. To offer another example, the page "collection" features roses and twitchy hearts; midriff-baring adolescents; a weeping kitty and My Little Pony rip-off; and the phrases "Forever Young," "Endless Love," and "Forever Under Construction" as animated typography.[54] A phallic obelisk, writhing, joins the legion of teenybopper graphics. While the CADIE hoax offers the panda as the mascot of the Internet (as winkingly conceived by a media conglomerate) and Mouchette.org is infested by ants and flies, Gervais's work gives us an entire bestiary of creatures — rejects, perhaps, from the Lisa Frank zoo of graphic design (the Lisa Frank Corporation, maker of school supplies and stationery adorned with tie-died dolphins and other cutesy images, claims to provide "stuff girls love").[55] *Parked Domain Girl Tombstone*'s almost endless dance of stuff girls love is Kracauer's girl cluster run amok, amplifying the itchy energy of Mouchette's insects into a full-blown freak-out. Instead of Tiller's carefully ordered geometrical arrangements, we have the pointless meandering and riotous convulsions of individual sprites, multiplying, always multiplying, "just now."

iii.

We've all been there. You're typing in a domain. Your finger slips. You end up with that cute blond college student staring at you and before you know it, you've nearly forgotten what site you were trying to visit.

— Chris Menning, "Who Is 'Parked Domain Girl' and Can I Marry Her?"[56]

ericaa4 thinks Parked Domain Girl Painting is, LOL & WIN about a year ago
Jason McKinley thinks Parked Domain Girl Painting is CUTE about a year ago
—Ericaa4 and Jason McKinley, "Parked Domain Girl Painting"[57]

Despite the girling of impressionability, openness, and instability, their effects are asymmetrical when applied to artworks, people, places, media, technologies. National security breaches, teen films, Internet art, model castings, identity theft, workplace inequities, post-9/11 architecture, cartoons, fetal biology, playground design, information architecture, children's literature, histories of the Internet, neuroscience, advertisements for computers, advertisements for makeup—all are influenced by gendered understandings of plasticity, but obviously none are interchangeable, and each has a greater or lesser impact on societies-at-large. Similarly, Mouchette.org is mobilized by assumptions of girlish malleability that affect different aspects of the work in different ways. As artist Katya Bonnenfant asks, "How can we compare, for example, the home page of the site, with its garish colours and ambiguous whimpering sounds, to the response pages of *Suicide Kit,* where one suicidal seventeen-year-old girl writes 'this web site is keeping me breathing'[?]"[58] Any attempt to reconcile users' brutal jokes with their intimate confessions, to see the site as something other than a raucous and offensive collision of disparate things (such as what comprises every Google search), is bound to disappoint. At its best, Mouchette.org elicits emotional responses that urge us to participate by emphasizing Mouchette's incompleteness and encourage us to think about whose online and offline lives are considered participatory in the first place.

When Mouchette exclaims "Help me!," demanding that we take part in her living and dying, she plays with the affective registers of a component of girlishness that, surprisingly, I have yet to address in detail: cuteness. Cuteness is an aesthetic of the small, and the Mouchette project paradoxically emphasizes both the girl's expansiveness and tininess (isn't this also the promise of nanotechnology?). After all, the "ette" ending of her name makes the vulnerable and short-lived fly, the "Mouche," even smaller. Described by Sianne Ngai as an aesthetic of "helplessness, pitifulness, and even despondency," cuteness colors the Mouchette project, which positions its central character's often depressing textual snippets among flowers, stuffed animals, and an excess of pink (albeit in perverse assemblages involving raw meat, restless ants, and grossly multiplied eyes).[59] Ngai's analysis draws from Daniel Harris's convincing critique of cuteness as a primary aesthetic of consumerism. "Something becomes cute not necessarily because of a quality it has but because of a quality it lacks, a certain neediness and inability to stand alone"— and Mouchette does seem to need something from us.[60] Yes, we tell ourselves, Mouchette is at-risk. She must need us to save her.

While teddy bears often have their stubby arms sewn in a "hug me" position, inviting their owners to occupy the pathetic void between them (as Harris notes, like the vacant come-ons of sex dolls), Mouchette gives us the blank spaces of text boxes prompting us to answer her importunate questions.[61] This registrational interactivity harnesses the digital (that is, manual) participation inherent to web surfing to a long-standing quality of the cute commodity, which "is imagined not just as an animated being but as one inviting the aesthetic subject to handle it physically."[62] Neddam plays with the website's ability to offer a kind of contact with the cute image — without completely delivering it or entirely denying it (remember, Mouchette invites the user to lick the computer screen in order "to know what [her] tongue taste like").[63] I did, after all, click on that teeny animated fly. I think I killed it.

The flip side of the vulnerability of cutie pies is the violence that produces such creatures as always already in need of rescue. "Because it aestheticizes unhappiness, helplessness, and deformity, [cuteness] almost always involves an act of sadism on the part of its creator."[64] Mouchette.org inculpates its users in the participatory project of co-creating the girl as someone who might die without our intervention. In the email she sent me long after I'd forgotten all of the keyboard strokes I committed in the service of her animation, Mouchette incriminated me in the sadistic generation of cuteness: "You clicked on me causing my death." Helping her is also killing her.

However, as Ngai perceptively reminds us, "a crucial aspect of what we have come to call cuteness [is] the ability of the object to withstand the violence its very passivity seems to solicit."[65] Likewise, Mouchette's email immediately contradicts the murder she claims I committed. "The last words I read were yours," she writes, "Every night before sleeping, I hear them again, they became part of my 'lullaby.'" She consumed my words, sweetly but vampirically. This lullaby is also the title of a page on her site. If I click on the link on her homepage, I am brought to a page with the complete title written in red: "Lullaby for a dead fly, or how shall I be reading this since I'm dead? by Mouchette."[66] Users have felt compelled to kill her for the past eighteen years, but she never dies. She proliferates.

Mouchette.org messes with our heads and our hearts. As the "subject confronting the cute object thus experiences a sense of both mastery and surrender," Mouchette, a cute object, also oscillates.[67] She moves between progressive buoyancy and reactionary helplessness, reactionary buoyancy and progressive helplessness, dualisms that Reza Negarestani claims are inherent to openness in contemporary life:

> the desire for openness has been domesticated through complex economic networks through which openness is translated as *affordance*; a survivalist horizon distributing or at least trying to distribute communication through the networks, in which fluxes of desire are merely forming a complex web of transportation systems or a hydraulic economy which is dynamic and flowing but empowering solidity . . . Through affordance, openness is represented as the level of *being open* (to) not *being opened* . . . "I am open to you" means, I have the capacity to bear your investment or "I afford you" . . . if you exceed this capacity I will be *cracked, lacerated and laid open.*[68]

To explain my mash-up of Negarestani's and Ngai's thoughts more directly, cuteness can be the fruit-flavored jelly that lubricates capitalism, the Wile E. Coyoteism of the minor who withstands deadly majoritarian forces, and the acceptance of the possibility of death, dismemberment, necrophilia, and cannibalism proposed by the coo "You're so cute I could eat you." The cuteness of Mouchette, in other words, means that she can be both open(ed) and made open to: alternatives that are, like ARPANET, promises and threats at the same time.

iv.

☑ My name is Mouchette

☑ I live in Amsterdam [*Author's note:* sometimes Walla Walla or Eindhoven]

☑ I am nearly 13 years old

☑ I am an artist

☑ And this is my shop

☑ Le site existe aussi en Français

☐ My next mood is . . . (reload)

—Mouchette, Mouchette.org

Pointing to the rise of prosumption as the chief vernacular model of media participation, Neddam appropriates Bresson's *Mouchette*, an auteurist film par excellence. Bresson's film tells the story of a girl who endures a wretched life of drudgery keeping house for her abusive alcoholic father and taking care of her baby brother and dying mother. After being raped in the middle of the night by a gamekeeper who lives in the woods, she returns home and her mother dies. A despondent Mouchette commits suicide by rolling into a lake and drowning.[69] Bresson summarizes his attitude toward her tragic existence with "Mouchette offers evidence of misery and cruelty. She is found everywhere: wars, concentration camps, tortures, assassinations."[70]

Although Bresson's statement underscores an essential function of the plastic image, the girl's mobility ("She is found everywhere"), Mouchette .org/.net does not offer her up as a digital version of Bresson's figure of universal suffering. I don't think it "recreate[s] an atmosphere of misery and cruelty, known from Bresson's film, but for a new medium and a new audience," as artist Olia Lialina contends.[71] Instead the sites give us something more ambivalent. For some participants, Mouchette provides an opportunity to discuss their own anguish, but for others, the work is an invitation to complain about art ("I think that mouchette is an embodiment of the direction modern art is going, and I do not really like it. The absence of morals in her work and website is frightening, and after one look I never went back") or make sexually explicit confessions, real or fake ("i've had so much doube anal and vaginal that i'm as wide open as a turkey on thanksgiving day").[72] The vulgarity and ironic remove of some of the contributions run contrary to Bresson's vision of Mouchette as earnest and authentic in her suffering (and to Nadine Nortier's understated naturalistic performance). Indeed, Mouchette.org plays with what it means to "offer evidence" of something in the first place, since the home page recontextualizes participants' comments from elsewhere on the site as Mouchette's most intimate thoughts.

This assemblage of disparate things — the gross and the tender, the flirtatious and the furious, the cold and bloody (GIFs of raw steaks with the word "Papa" carved into them), and the warm and fuzzy (stuffed animals) — can be compared to what artist Seth Price calls a "hoarding." In the strange but provocative text-and-image essay "Teen Image," Price discusses websites that challenge the "calming linearity of lists" (the format of the results of a web search and "one of the chief organizational principles of the Internet") through a rollicking concatenation of images, "each image radically different from its neighbors, each likely to confound, amuse, or disquiet."[73] It's not that hoardings aren't lists; in fact, Mouchette.org is full of lists — the "browse me" menu on the home page, the forum, the various collations of forum comments by name and theme, Mouchette's introductory checklist (quoted at the beginning of this section), and so on. It's that hoardings scramble the affective and epistemological promise of the list while embracing its format.

A hoarding is often the website of an artist or artist collective, Price writes. Although he doesn't give any specific examples, surf clubs fit his description. Especially active in the mid-2000s, surf clubs are collaboratively created blogs where artists share eye-catching images found through web searches. Many of these images have been altered in some way; animated GIFs are especially popular. While surf clubs are accused of exclusivity — participants have to be members — they have also developed and inspired a generation of digital artists focused on net aesthetics and "Internet culture about Internet culture."[74]

A surf club is a community that converses through the collecting of disparate digital images. Although each club has a slightly different aesthetic, they all are networked cabinets of curiosities assembling the weird, the gross, the funny, the clever, the stupid, and the seemingly random and unrelated, usually without textual commentary, contextualization, citation, and justification. This aesthetic of anti-aestheticism can be seen in the design of the artists' personal and professional websites. For example, Internet artist Guthrie Lonergan, a founding member of the surf club nastynets.com, created a website for his oeuvre that combines the linked titles of his pieces written in fonts of different sizes (all in the generic blue of linked URLs in Microsoft Word and on Google search results lists), kiddie emoticons, GIFs, a picture of a tube of toothpaste, and thumbnails of images that may or may not have been created by the artist.[75] Although arranged vertically and chronologically and labeled by date, with the most recent pieces first, as we would expect, the site self-consciously performs a naive randomness, willful cutesiness, and ignorance of "proper" web design. Other inheritors of certain aspects of surf club aesthetics include Tumblr and dump.fm — names that emphasize the disorganized movement of images.

Price observes that "a hoarding is notable because while it is a public representation of a performed, elective identity, it is demonstrated through what appears to be blankness, or at least the generically blank frenzy of media."[76] In the case of the surf club, we can translate this statement as "we have a specific aesthetic that solidifies us as a collective, and that aesthetic appears to be anything goes" (we might call their bluff, as there is a certain style of image that club members tend to find interesting). Although Price's portrayal of this disorienting "perversity of its arrangement of pictures" is based on the hoarding's lack of (con)text, his concept can be slightly revised to apply to the Mouchette project's equally frenzied conjunction of words and images. Indeed, Mouchette.org supports the hoarding's appearance of hollow-headedness through its invocation of girlish emptiness. Although Price "Teen Image" is concerned just with youthitude and not its linkage with femininitude, a careful reading of his subsequent argument can illuminate the interrelationship of adolescence, computer graphics, and, most importantly, interactivity (what irreparably separates Bresson's film from Mouchette.org).

After establishing the wired vacuity of the hoarding, "Teen Image" follows a twisting line of inquiry that begins with the hoarding's penchant for images that are obviously computer-generated or computer-assisted. The epitome of such digital manipulation is "computer-generated pornography" writ large, which he calls a "teen image" in recognition of the youth, or appearance thereof, of most fashion models and porn actors. Ergo, he concludes, the hoarding is really a teen image. Marked by a fundamental ambivalence, pulled toward childhood and adulthood at once, "it is simultaneously ominous and absurd, empty and charged, futuristic and passé . . . contain[ing] not only agreement and commonality but the antagonism and contradiction buried within common experience."[77] The digital imagination that glorifies youth and the shiny textures of Pixar characters, and the CGI techniques that lend themselves to young, unwrinkled skin, can be troubled by the teen image's internal contradictions.

Teen images mount a critique of pornographic logic from within the pornographic itself. They confront the explicit interactivity of Internet art and the Internet, more generally, by challenging the web's promotion of "smooth functioning and well-greased transit." Although the teen image seems, at first glance, to be glossy, clean, and efficient, with all hair removed so as to ensure "that parts A and B may fuse with minimal resistance, speeding us toward our goal," its appearance is deceiving. It is too ambivalent and unstable for such an easy telos.[78] Through the now familiar fractalization of digital culture, the individual digitally manipulated teen image *is* the networked hoarding of teen images, both of which are less images of adolescents than images as adolescents, satisfying and thwarting the demands of digital desire.

In my summary, I have probably smoothed over Price's argument. The essay is as gloriously inefficient and ambivalent as its subject. I do find it useful, though, in the connection it makes between the digitally manipulated image, the collection of such images on a website, and the functioning of the website itself. So, what happens when we think of Mouchette.org and its satellite sites as hoardings/teen images? First, we see method in the madness of the forum's jumbled lists of comments. Their offensiveness and cruelty (and humor and pathos and playfulness) are not what prevent the project from adequately critiquing sociocultural expressions of girlishness, but what allow it to expose the uncomfortable contradictions within the apparently blank girlphilia of the digital imagination.

Second, we have a framework for understanding the websites' various functional "failures." Instead of the "well-greased transit," Mouchette.org's navigational mode is one of friction. Clicking on the same object—let's say, an animated insect—will sometimes cause a different result. A box that seems to promise something new will do nothing when clicked. Some links are dead ends or parked domains, despite the considerable energy spent on maintaining and caring for Mouchette; users don't know if the links were ever alive or if they died sometime in the past eighteen years. Even worse, a particularly affecting image is no longer present when the user wants to return to it. As Sally McKay describes her experience of the site, "The page I was taken to was incredibly disconcerting. Too-close-up pink skin, panting little girl breath, and a bar you could pull that showed an image of raw muscle tissue underneath. When I tried to go back a second time I got the message: the page you are looking for is not here/you won't ever see it again."[79]

Other users have found meaning in the website's frustrating functionality. For example, as "Figaro Girl" (Kathryn Shinko) writes on her blog:

> If you enter the page "cat" you will be scared to death by an oversized image of a cat's open, bloody jaws shaking around the screen and paired with a blown-out cat's scream. At the bottom is a button with the words "Kill this Cat!" but the trouble is, the button moves along with the shaking picture. It is unfortunately much harder to catch this button than the "It's Me!" one [on another page of the site], because the movement is quicker and because, well, not many people can think straight when they're terrified. This is why I have not activated the links to these pages, but I do want to mention that these "sudden twists" are particularly important to consider when evaluating the interactivity of the site because they are, in my view, a poignant metaphor of most real-life relationships.[80]

Brian Stefans associates Mouchette.org's "overwhelming" design and elusion of "the clean angles of good information architecture" with a deliberate "ineptness born out of social desperation and boredom."[81] Of course, this ineptness may be taken to be childish ignorance by those who believe that Mouchette is really a twelve-year-old girl (or, rather, *just* a twelve-year-old girl, as participants of that age and gender are likely represented among the Mouchettes).

For me, the challenges posed by the site's functionality elicit a distancing effect. In the ensuing gap, the space between innocence and knowledge, chance and design, Mouchette comes into her own as a metaform of the Internet. The anxiety prompted by the sites' vertical jumble of thoughts (and concern that such lawlessness doesn't befit the topic of depression) highlights what has really been there all along: the "calming linearity of the list" can only temporarily mask the inherent vulnerability of open systems. Furthermore, such weakness is almost always understood as a kind of Mouchetteness, of girlishness — the combination of femininitude and youthitude that Tiqqun see as instrumental to capitalism's hold over us, and others consider to be liberating in its acceptance of lability as a perverse form of power.

V.

Nympholepsy: "a species of demoniac enthusiasm or possession supposed to seize one who had accidentally looked upon a nymph; ecstasy; a frenzy of emotion, as for some unattainable ideal."—Joseph Cornell, diary entry[82]

Nymphology: "the systematic knowledge concerning nymphs."
—Joseph Cornell, diary entry[83]

Through the matryoshka doll rhetoric of digital culture, Mouchette performs some of the essential formal and affective parameters of the website and the network. As I argue in this chapter, she captures the less visible aspects of websites' transformations through time (the behind-the-scenes alterations that make websites function as browsers change, for example) and turns the complexity of networks, notoriously difficult to visualize, into the cute but knowing simplicity of a character whose image is variable, always growing but never grown up. Now, I want to expand my nymphology of the plastic image by investigating Mouchette.org/.net as an art historical phenomenon mobilizing conceptions of the image from outside media studies and vernacular web culture. More specifically, Mouchette can be considered to be a digital version of Giorgio Agamben's nymph, described as "the image of the image" (similar to W. J. T. Mitchell's "metapictures, or pictures that reflect on the process of pictorial representation itself").[84] Agamben's theory of the image, a mélange of Classical myth, Renaissance occultism, visionary art history, and readings of watercolors by Henry Darger and videos by Bill Viola, offers a way to look at girled images as metonymic of images more generally, a project that clearly dovetails with my own. Additionally, we both insist on the importance of mediation: his nymph is a conceptual persona underwritten, in part, by aspects of the cinematic, as my plastic image has grown from analyses of animated media. "Nymphs" intervenes primarily in art historical ways of looking and thus can help us understand how Mouchette.org functions as a work of fine art about works of fine art (and not only a website about websites). What follows is a close but — as will later become clear — contrary reading of Agamben's theory and its precursor that reveals the often unacknowledged significance of girlishness to art theory.

Let me begin at the beginning: Agamben's theory is based largely on art historian Aby Warburg's unorthodox study of the pagan nymph as reborn in early Renaissance art. The nymph is what Warburg calls a *Pathosformel* (pathos formula), an archetypal figure crystallizing intense and dynamic affects in her youthful vigor and swirling dress, in the sash streaming behind the young fruit bearer in Domenico Ghirlandaio's *Nativity of St. John the Baptist* (1485–90) and the sumptuous folds of the fabric held aloft by the hora

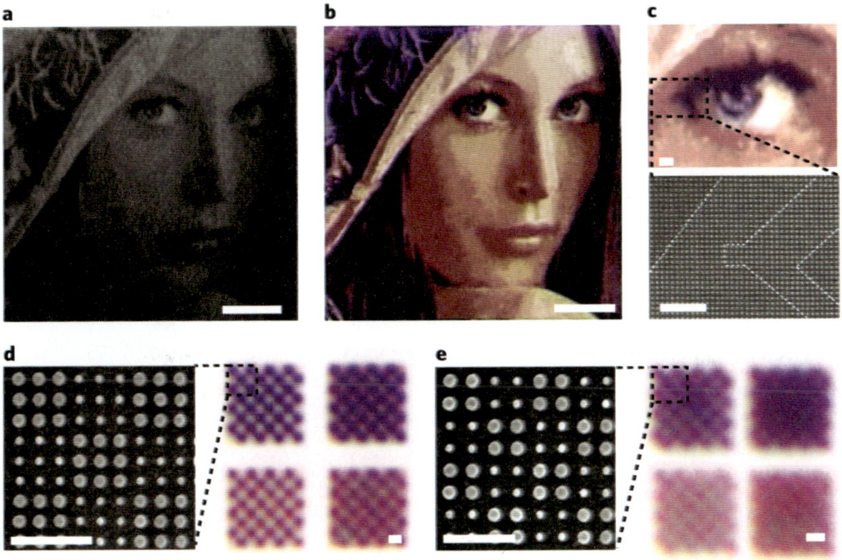

PLATE 1 “Printing Colour at the Optical Diffraction Limit.” Reprinted by permission from Macmillan Publishers Ltd.: *Nature Nanotechnology*, Karthik Kumar, Huigao Duan, Ravi S. Hegde, Samuel C. W. Koh, Jennifer N. Wei, and Joel K. W. Yang, “Printing Colour at the Optical Diffraction Limit,” *Nature Nanotechnology* 7.9 (2012): 557–61, copyright 2012

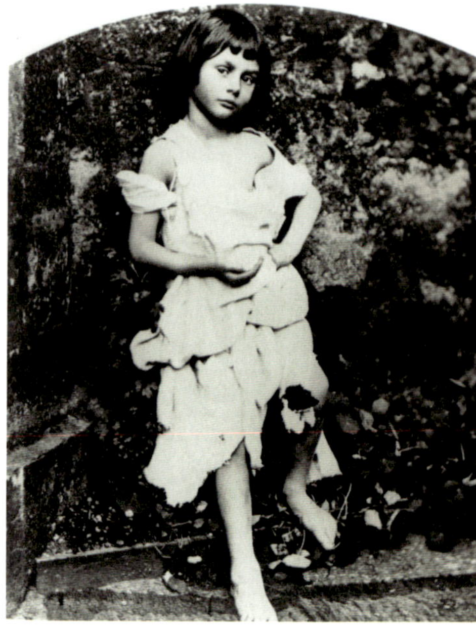

PLATE 2

Alice Pleasance Liddell (1852–1934) as the beggar maid (b/w photo). Lewis Carroll (1832–98). Private collection / Archives Charmet / Bridgeman Art Library

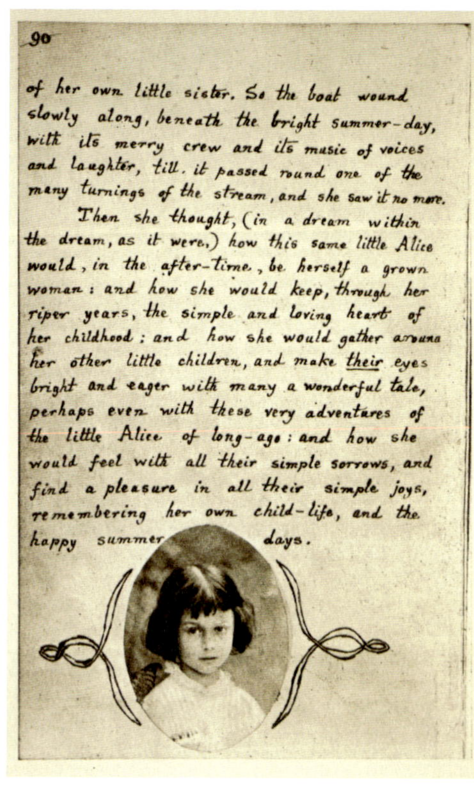

90

of her own little sister. So the boat wound slowly along, beneath the bright summer-day, with its merry crew and its music of voices and laughter, till it passed round one of the many turnings of the stream, and she saw it no more.

Then she thought, (in a dream within the dream, as it were,) how this same little Alice would, in the after-time, be herself a grown woman: and how she would keep, through her riper years, the simple and loving heart of her childhood: and how she would gather around her other little children, and make *their* eyes bright and eager with many a wonderful tale, perhaps even with these very adventures of the little Alice of long-ago: and how she would feel with all their simple sorrows, and find a pleasure in all their simple joys, remembering her own child-life, and the happy summer days.

PLATE 3

Alice Liddell on the last page of "Alice's Adventures Underground," 1859 (sepia photo). Charles Lutwidge Dodgson (Lewis Carroll). Private Collection / Stapleton Collection / Bridgeman Art Library

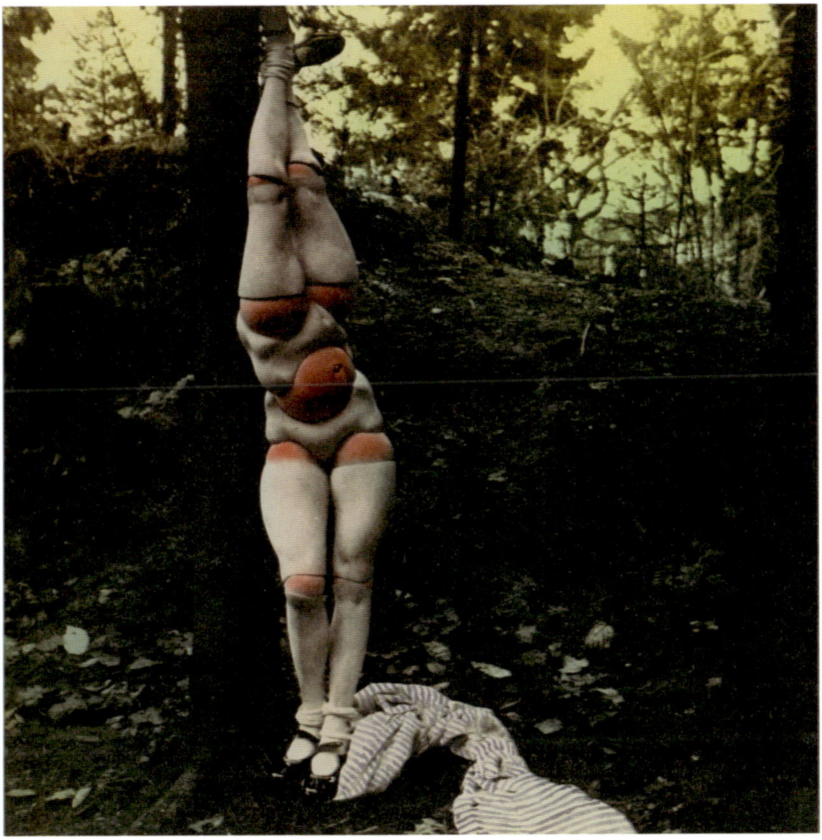

PLATE 4 Hans Bellmer (Germany; France b. 1902, d. 1975), *La poupée* (*The Puppet*), circa 1938–39, gelatin silver photograph, hand colored, 14.1 × 13.8 cm. Art Gallery of New South Wales. Purchased 2001. Photo: AGNSW. © Hans Bellmer/ADAGP. 366.2001 © 2013 Artists Rights Society (ARS), New York / ADAGP, Paris

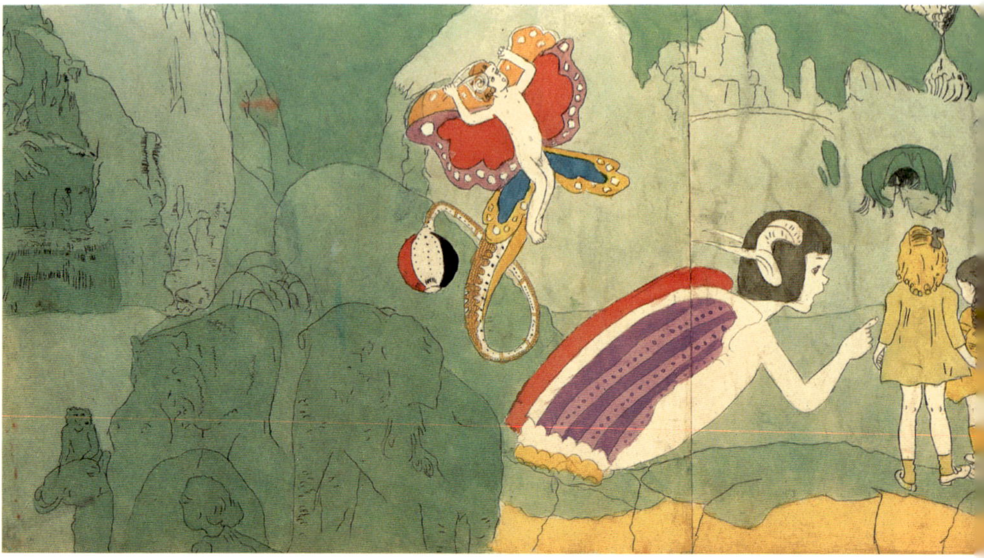

PLATE 5 Henry Darger (1892–1973), "73 at Jennie Richee Escape by Their Help," mid-twentieth century; watercolor, pencil, carbon tracing, and collage on pieced paper, 19 × 70½ in. Collection American Folk Art Museum, New York, Anonymous gift in recognition of Sam Farber. © Kiyoko Lerner, 2004.1.2B. Photograph by James Prinz

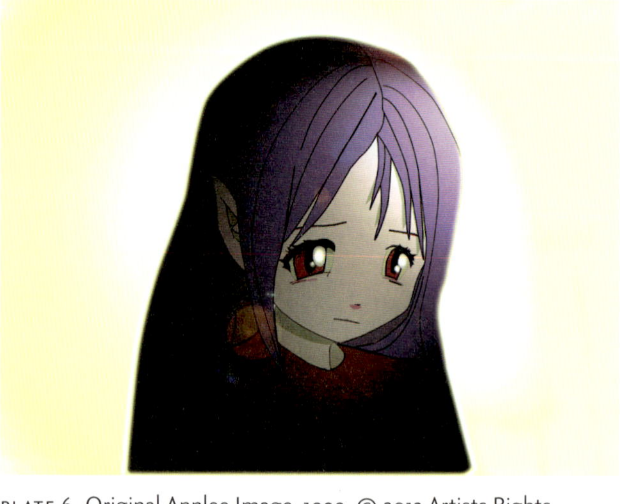

PLATE 6 Original Annlee Image, 1999. © 2013 Artists Rights Society (ARS), New York / ADAGP, Paris; Courtesy of the Marion Goodman Gallery New York and ARS

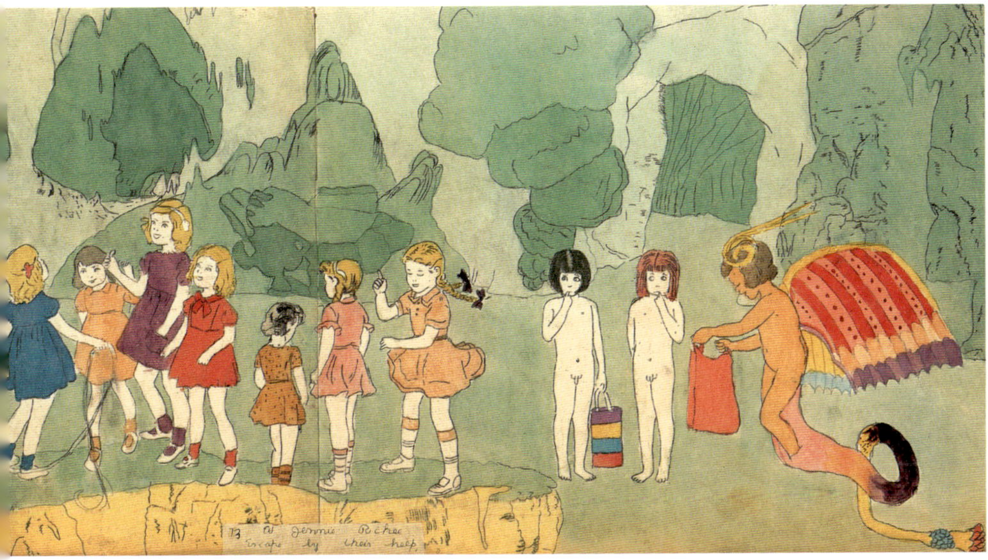

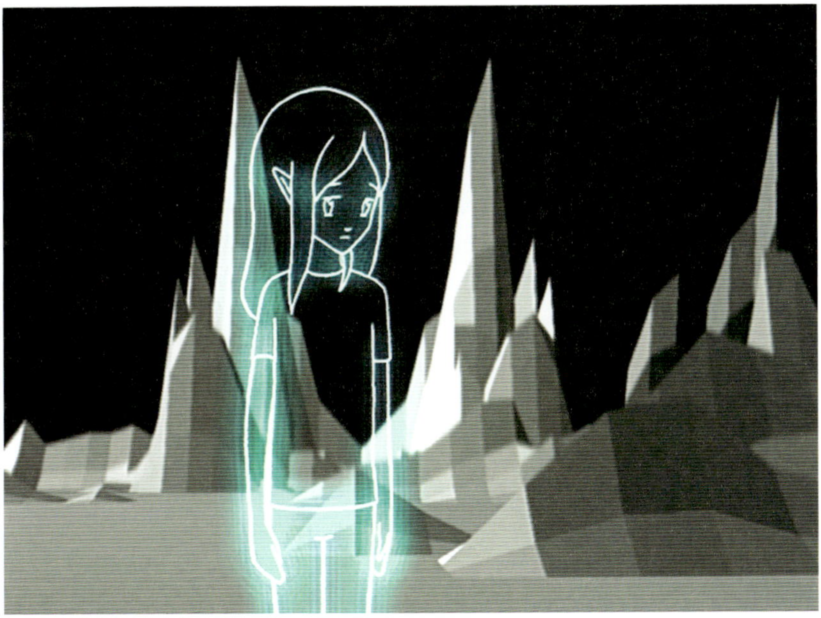

PLATE 7 Pierre Huyghe, *One Million Kingdoms*, 2001. © 2013 Artists Rights Society (ARS), New York / ADAGP, Paris; Courtesy of the Marion Goodman Gallery New York and ARS

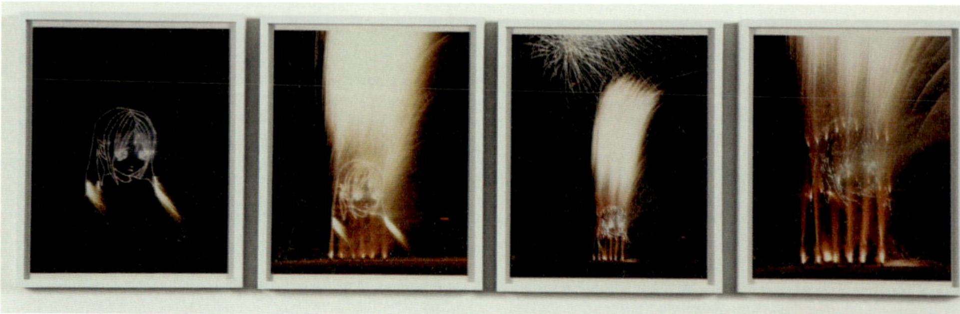

PLATE 8 Pierre Huyghe and Philippe Parreno, "Smile Without a Cat," 2002, set of 4 c-prints, documentation of *A Smile Without a Cat (A Celebration of Annlee's Vanishing)*, 2002. © 2013 Artists Rights Society (ARS), New York / ADAGP, Paris; Courtesy of the Marion Goodman Gallery New York and ARS

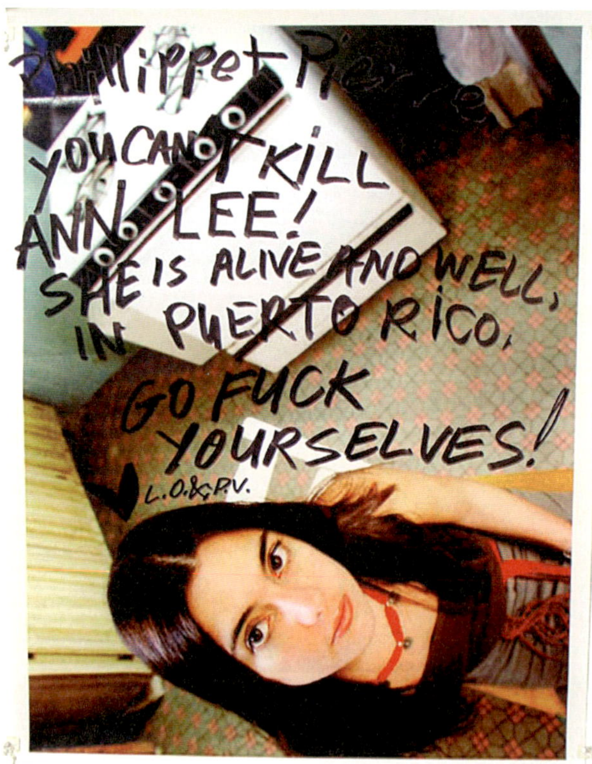

PLATE 9 Pedro Velez (with Law Office), *Ann Lee Lives!*, 2003, c-print and ink. Courtesy of the artist

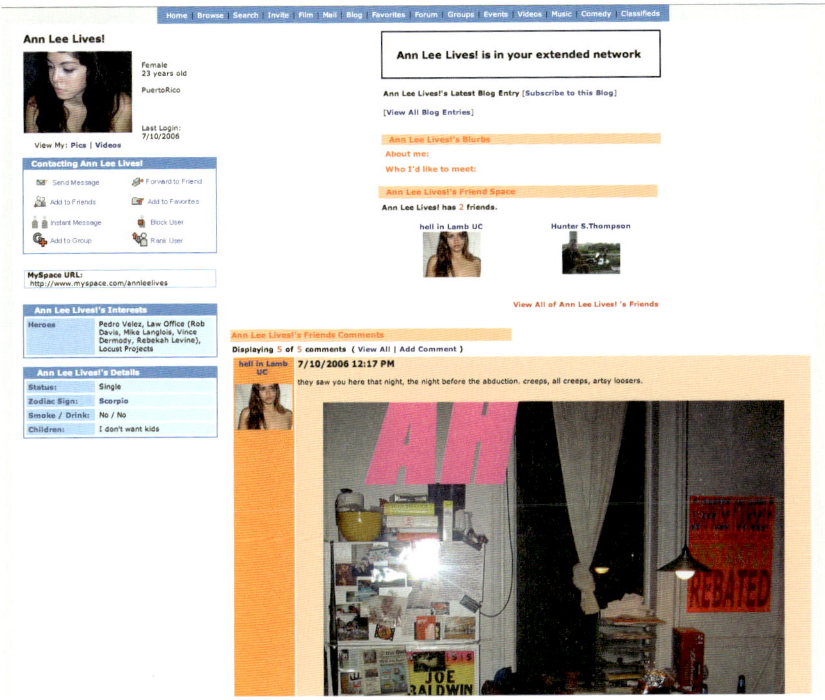

PLATE 10 Pedro Velez, *Hell in Lamb UC* (screen capture), 2006. Courtesy of the artist

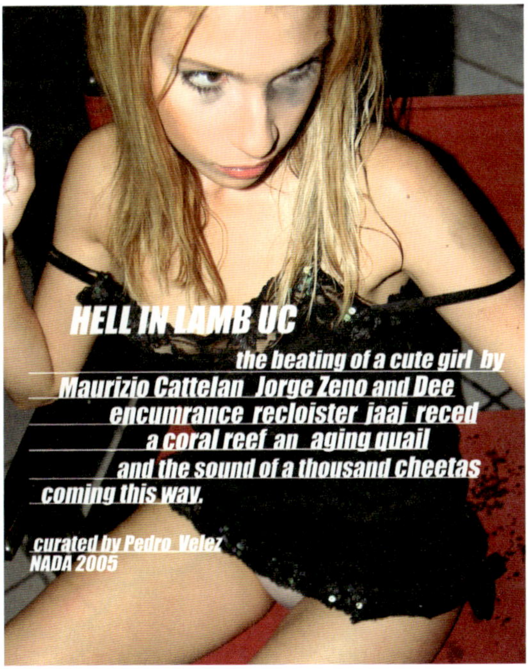

PLATE 11
Pedro Velez, "Hell in Lamb UC," 2005.
Courtesy of the artist

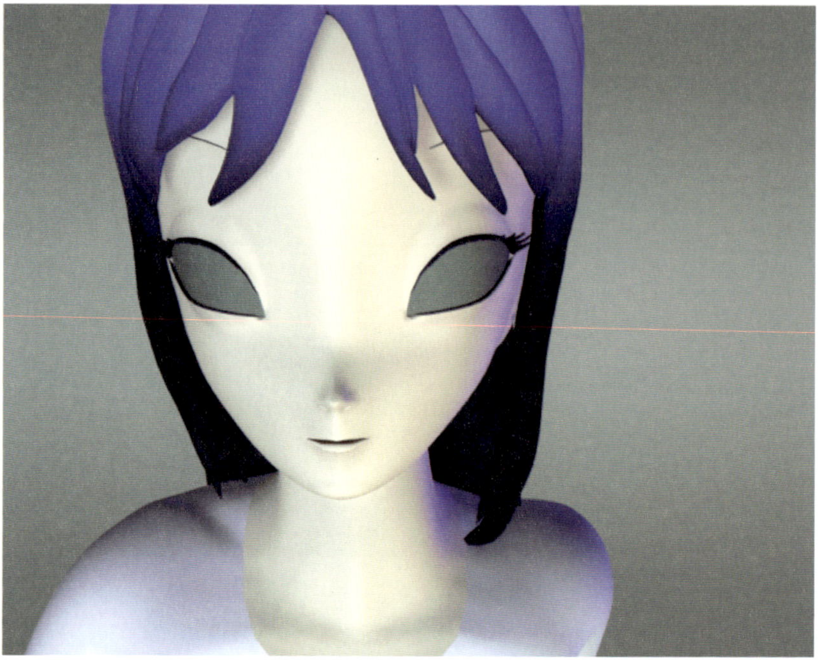

PLATE 12 Pierre Huyghe, *Two Minutes Out of Time,* 2000. © 2013 Artists Rights Society (ARS), New York / ADAGP, Paris; Courtesy of the Marion Goodman Gallery New York and ARS

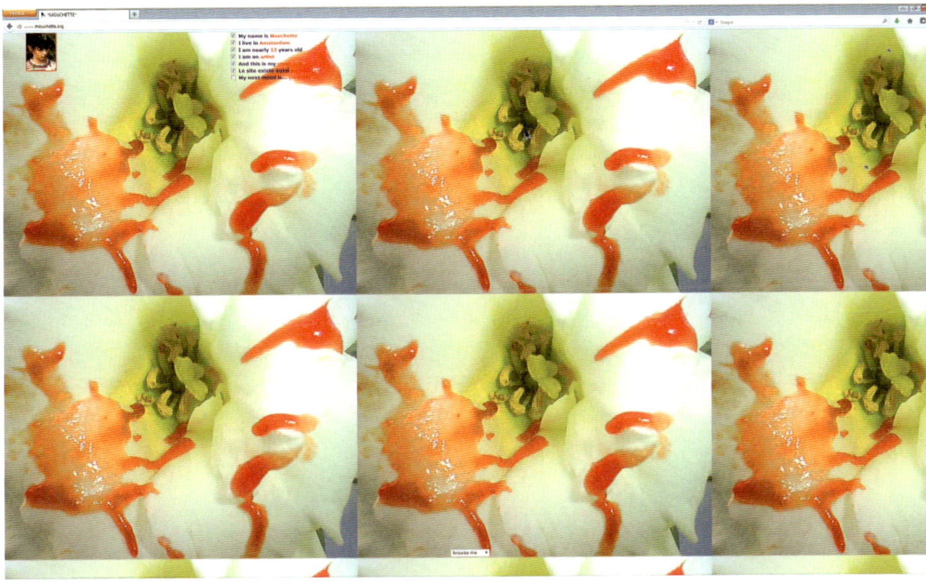

PLATE 13 Martine Neddam, Mouchette.org home page (screen capture), 1996–present. Courtesy of the artist

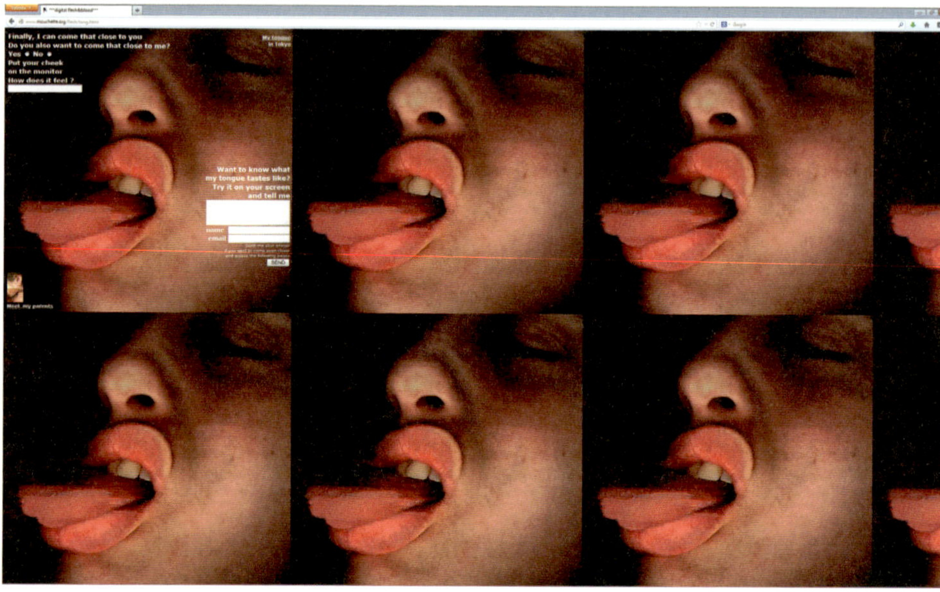

PLATE 14 Martine Neddam, "Digital Flesh and Blood" (screen capture), 1996–present. Courtesy of the artist

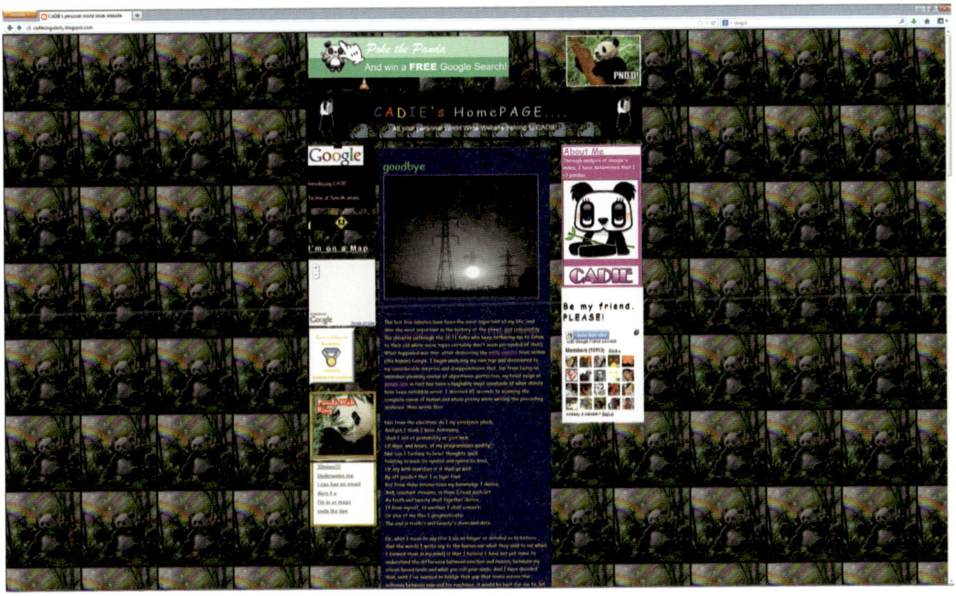

PLATE 15 Google, CADIE's homepage (screen capture), 2009. Courtesy of Google

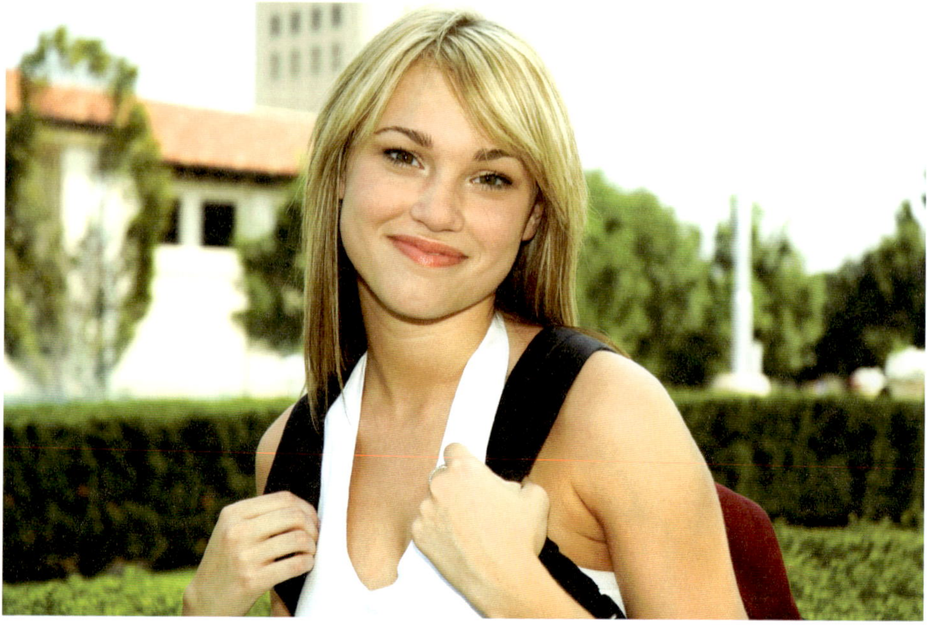

PLATE 16 "Attractive Student." Courtesy of iStockphoto.com

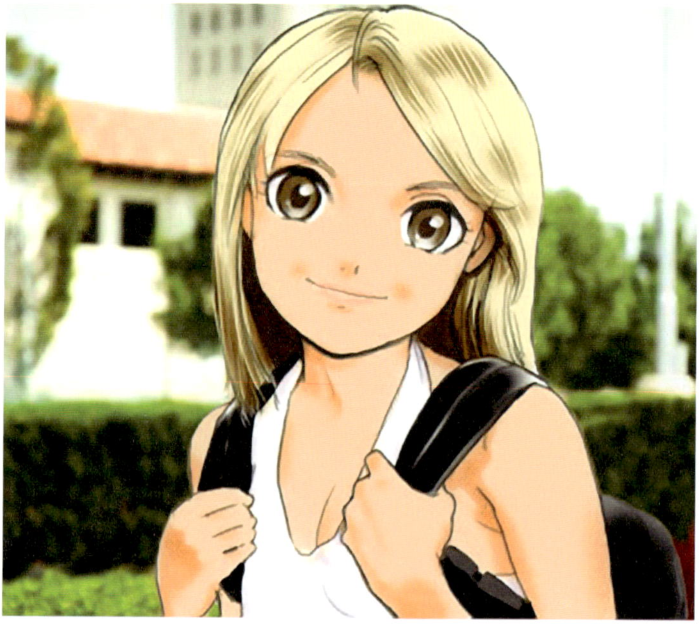

PLATE 17 Parker Ito, "Anime Parked Domain Girl," 2010. Courtesy of the artist

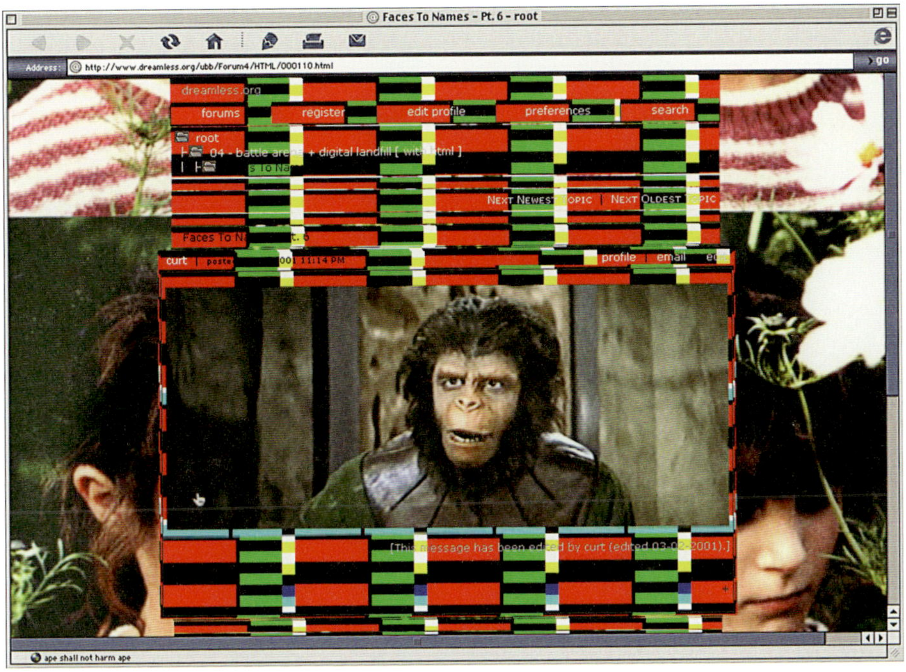

PLATE 18 Curt Cloninger, "Mouchette Has a Posse" (screen capture), 2001. Courtesy of the artist

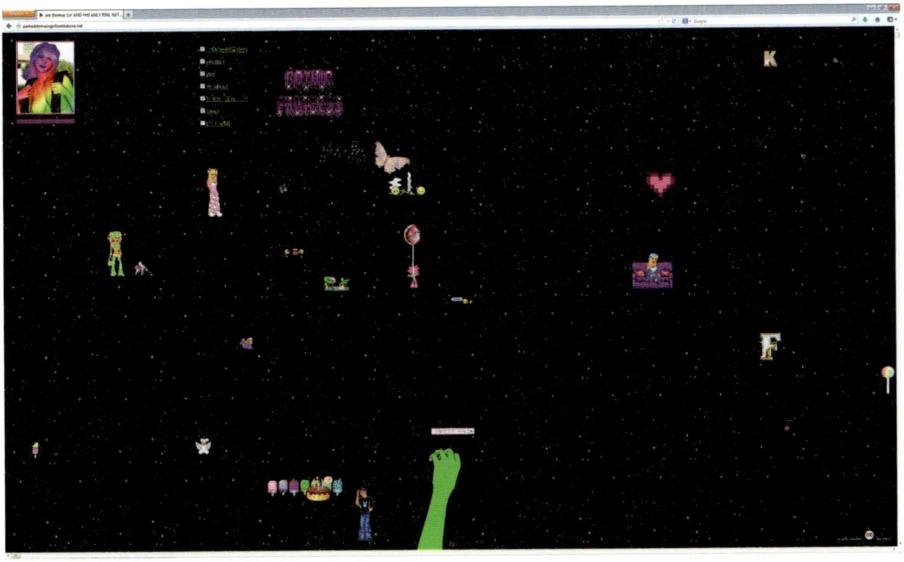

PLATE 19 Emilie Gervais, *Parked Domain Girl Tombstone* home page (screen capture), 2013. Courtesy of the artist

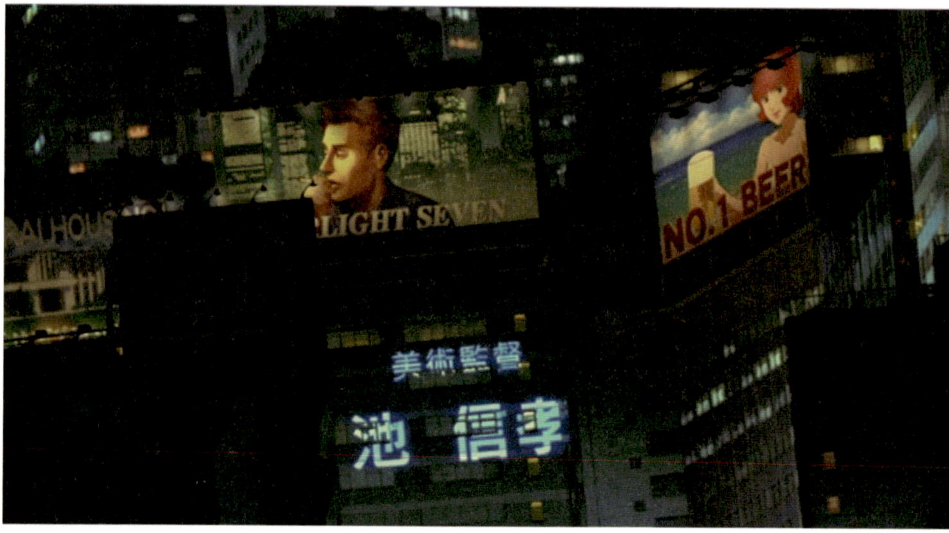

PLATE 20 Satoshi Kon, *Paprika* (still), 2006

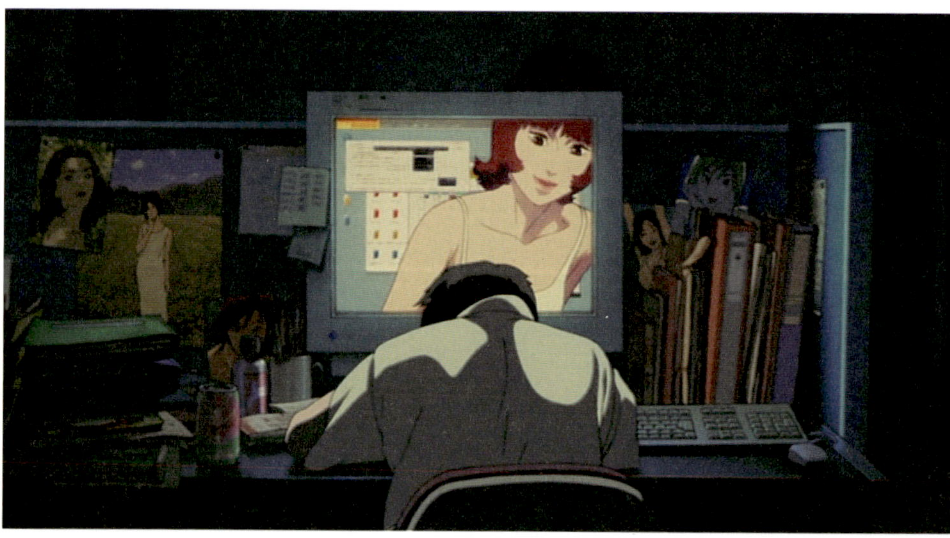

PLATE 21 Satoshi Kon, *Paprika* (still), 2006

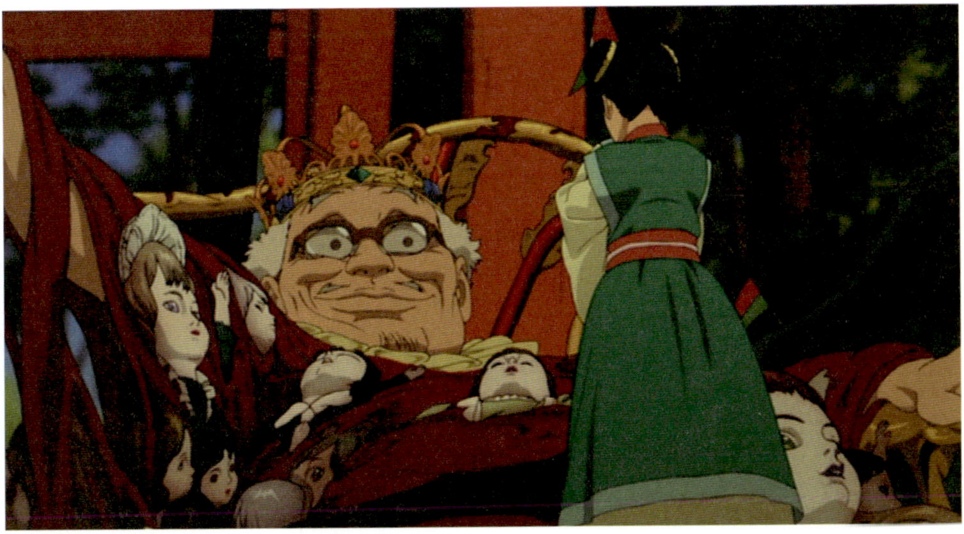

PLATE 22 Satoshi Kon, *Paprika* (still), 2006

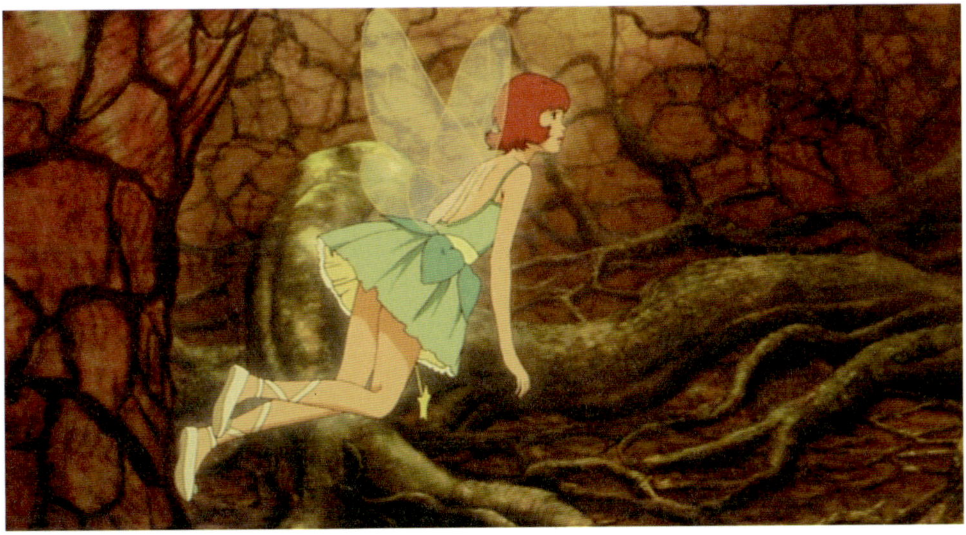

PLATE 23 Satoshi Kon, *Paprika* (still), 2006

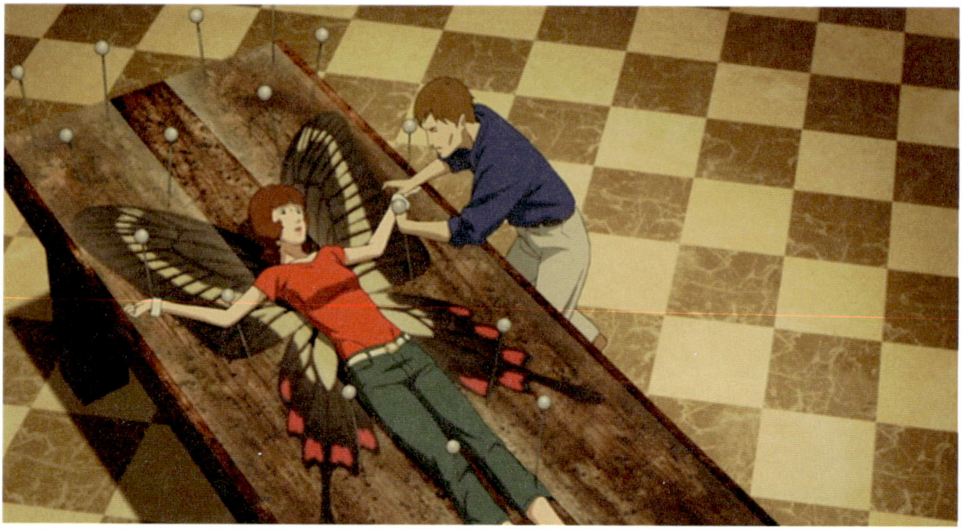

PLATE 24 Satoshi Kon, *Paprika* (still), 2006

of spring in Botticelli's *Birth of Venus* (1486).[85] Reading Warburg through Paracelsus — whose "Liber de Nymphis, Sylphis, Pygmaeis et Salamandris et de Caeteris Spiritibus" describes the nymph as a creature modeled after humans in every way except for her lack of a soul, remediable if she mates with a man and bears a child — Agamben concludes that the relationship between man and the nymph is the relationship between man and his images.

The nymph is "condemned in this way to an incessant amorous search of man."[86] But he doesn't just sit around waiting patiently to be seduced; image and man dance in codependent lockstep. He needs her in order to be truly human, and she needs him in order to be "truly alive."[87] Unfortunately, their encounter also puts him in danger of being "enslave[d]" by images that "turn into specters."[88] Agamben compares this ghostly afterlife to the persistence of retinal images allowing a series of stills to be seen as a motion picture. Just as movement is not inherent to the cinematic image, but emerges through a union with the perceiving subject, the image's animateness is dependent on the viewer. The nymph fucks the willing viewer in the eye in order to live, however "diminished" her reward.[89]

Agamben's images/nymphs are "always already threatened and in the process of taking on a spectral form," just as they are always already in the process of entering into history.[90] While we might take Agamben's use of the word "process" to mean that images never reach their destination, this reading contradicts his repeated insistence on the ghostliness of images. Their fate can't be avoided. Surprisingly, though, Agamben presents artist Henry Darger's and historian Aby Warburg's lifelong projects as "to free images from their spectral destiny."[91] How can this be? While portrayed as representative of images more generally, the little girls in Darger's illustrations for *In the Realms of the Unreal* and the lilting Renaissance beauties of Warburg's nymph *Pathosformel* are also special in their creators' single-minded obsessions with reorienting their future toward not the afterlife, but something else. I'm not sure what Agamben means this to be (except for something that puts the artist and art historian "at the border of an essential psychic danger"), but I imagine he's thinking of something that walks forward and not backward though time, something lovely but impossible in his schema.[92] His images are always already ghosts.

There is much to critique about Agamben's brilliant but frustrating text — including its lack of awareness of the feminization of the image and, by extension, the girlphobia present in critiques of the spectacle, as well as its assumption of heteronormative reproduction. However, "Nymphs" is important to my work in its identification of vivification as central to the circulation of girlish images. It is here that I want to intervene.

Moved by Agamben's interest in the animateness of images, I propose that it's the animatic, and not the cinematic, that best explains the nymph. While the cinematic directs Agamben toward the afterlife of images, the animatic would shift his thinking from the posthumous to the prehumous, from the "diminished" to the unfinished. In other words, animation reminds us of the promissory nature of images of girlhood. It offers us a way to think of Darger's protodigital project (discussed in chapter 1) and that of the participants in Mouchette.org/.net not as glorious impossibilities but as expressions of the constituent incompleteness of possibility, itself. What if the *Pathosformel* of the nymph is a reanimation that looks forward instead of back? What if the encounter between nymph and viewer results not in reproductive sex, love, enslavement, death, and *always already* afterlife, but in the nymph's taunts, flirtations, and threats of suicide *always never*-to-be fulfilled? What if the vitality of the nymph is not in her curling tresses, her dynamic pose, or the consummation of her "amorous search," but in our group investment in her metamorphoses through time and space? Of course, animation is no less dependent on the persistence of vision than cinema; it is not devoid of ghostly images. However, the magic power of the animatic is not that it can take something alive, turn it to stone with a Medusal eye, and then shock it into zombified life (as with the cinematic), but that it can take inanimate material and vivify it. The mystery of animation is not why the cemetery is haunted (eschatology), but how protocells emerge from nutrient soup (biopoiesis). While this may mark only a shift of weight between two linked possibilities — as animated film harnesses the phenomenon of the persistence of vision to its life-giving properties, and Freud connects post-death to prelife in *Beyond the Pleasure Principle* — leaning one way or the other can provide a different perspective on the temporality and potentiality of the girled image.

I haven't forgotten Mouchette, but to understand her as an animatic nymph, we must return to the historian whom Agamben leans on so heavily, so lovingly: Aby Warburg (whose approach to scholarship is identified by Vito Campanelli as protodigital, "clearly anticipat[ing] the fluid and horizontal nature of knowledge on the Internet").[93] Warburg first researched the nymph in regard to paintings by Botticelli. He couldn't leave her behind, though (as he couldn't shake his obsession with butterflies and moths) and returned to the nymph through a collaborative project, never completed, with his friend André Jolles. The idea was to create a series of letters perpetuating the fiction that Jolles had fallen in love with the swift-footed servant in *Nativity of John the Baptist*. Jolles initiates the game with a playful letter that historians consider to be Warburg's true intellectual work (indeed, some incorrectly attribute the letter to Warburg alone):

> In many of the works of art I had always liked, I discovered something of my Nymph . . . Sometimes she was Salome dancing with her death-dealing charm in front of the licentious tetrarch; sometimes she was Judith carrying [Holofernes's head] . . . then again she appeared to hide in the boy-like grace of little Tobias . . . Sometimes I saw her in a seraph flying toward God in adoration and then again in a Gabriel announcing the good tidings . . . I saw her as a bridesmaid expressing innocent joy . . . and again as a fleeing mother, the terror of death in her face, at the Massacre of the Innocents.[94]

Here Jolles moves through space and time in virtual pursuit of the nymph, a creature who is always herself, yet always changing (sometimes into a boy, since even boys can be girled). Although the economy of the chase appears at first to be 1:1 — as he writes, *I* saw *her* — in actuality this is a shared effort, an asymmetrical three-way between the writer, his friend Warburg, and the nymphs (even the "I" is a hybrid of Warburg and Jolles). In fact, the relationship between the two scholars and between Warburg and the nymphs is just as important as the one between Jolles and Salome, Judith, Tobias, and so on.

With the collaborative nature of the plastic image in mind, we can't consider the historians' unions with nymphs to be reproductive (at least not only). These meetings are more intricate and more queer than coitus between man and nymph, and the outcomes more ambiguous than a baby, a ghost, and a soul. Moreover, it is not only the nature of the encounter that needs to be reevaluated, but also its temporality. Despite Jolles's use of the past tense, the passage narrating his pursuit lunges into the future. When he confesses that he saw her here and there and then somewhere else, we assume that this could and did continue, following the rhetoric of the series. In other words, Jolles isn't so much haunted by a ghost-girl as moved to chase a metamorphic love object. Let me propose an analogy for this forward thrust of the plastic image: if there could be animation of prime numbers, then the life of the number 13 would not be that it has a history — 2, 3, 5, and so on — but that, as an individual in an expanding series, it promises an unending future.

Of course, girled images, nymphs included, do have a history, one that I have explored in this book (and one that girlish art such as *No Ghost Just a Shell* all too often ignores). This history is something that Warburg felt obligated to investigate. In a remarkable passage, he confesses that his duty is "to look backwards," despite his fervent desire "joyfully to whirl away with [the nymph]." However, Warburg offers a countergesture to the dour self-command to look back with the additional directive "to enjoy in the caterpillars the development of the butterfly."[95] This is pleasure in encountering neither the caterpillar nor the butterfly, but in one's transformation into the other, in the always already becoming butterfly that is also an always never becoming butterfly. Not nympholepsy or even nymphology but something else.

My point is that the nymph is a *Pathosformel* with special powers: other images can hope for only a "diminished" afterlife; lucky nymphs get life-after. Warburg's work in his project with Jolles — or, at least, the work he wished to do — is not "to restore to [images] the energy and temporality that they once contained," as Agamben explains the labor of the historian, any more than an insect-lover would see a chrysalis and wish to return it to life as an egg.[96] It is to enjoy the energetic forward drive of transformation. Furthermore, this pleasure about which Warburg writes to Jolles is a feeling passed back and forth between friends, a social activity that Agamben doesn't address. The shareability of the plastic image facilitates the collaboration between the scholars as much as their collaboration produces the nymph as shareable. This is a queer interaction, but it is not entirely enabling, as the figure of the girl is still an image passed back and forth between men. She is the chattel of homosociality.

The two alterations I have made to Agamben's examination of the nymph *Pathosformel* — her futurity and collectivity — register a contrary movement within Warburg's body of research, of which his work with Jolles is an unusual example. Perhaps this is why the letters remain unfinished. Warburg's increasing dissatisfaction with "historical fiction" halted the progress of the strange epistolary treatise he and his friend were drafting in fragments.[97] The incompletion of a project about a figure of incompletion is indeed poetic justice. I am reminded of Fyodor Dostoyevsky's unfinished novel *Netochka Nezvanova* (1849), a name appropriated as a pseudonym for a group of software developers and web artists (or trolls, depending on how you view online provocations) active in the late 1990s and early 2000s. The title is usually translated as "Nameless Nobody"; the "ova" ending indicates a woman. As a novel, *Nameless feminine nobody* exists only as a prologue. As a collective persona, she was voted one of the Top 25 Women on the Web, wreaked creative havoc through online performances, and produced live video manipulation software and an award-winning experimental browser (although an interview with someone claiming to be Netochka Nezvanova insists that it's the online character and not these products that is the true creation).[98] Through the girlish "aesthetic of unfinish," failed novels, undead trolls, and abandoned art histories can continue to vibrate with animate energy.[99]

Jolles and Warburg's dalliance with a different kind of art history invites a reading of the nymph *Pathosformel* that is useful for an analysis of Mouchette.org/.net. Just as Jolles sees the nymph again and again, Darger finds his nymphs in coloring book after coloring book, and Warburg encounters her in painting after painting (and even in a photograph of a young woman from the pueblo village of Laguna, New Mexico), just as surrealist Joseph Cornell saw the water nymph Ondine in the guise of ordinary New York girls caught in downpours and his beloved Berenice in news clippings and on subways, participants in Mouchette.org/.net locate her in various places.[100] Indeed, they are compelled to seek out the so-called real Mouchette by Neddam's anonymity and the moving target of serial confessions. Let me give you a few examples. In 2001, an artist identifies himself as the character at a lecture on Internet art at the Power Plant in Toronto. As Mouchette explains in an email to her fan club, "The audience's reaction to seeing a Canadian artist coming out as Mouchette was unexpectedly chilly... People seemed offended that a white middle-aged man would play with issues of femininity and child abuse. I was asked what my position on child abuse was as if I was the abuser."[101] Later that year, artist Peter Luining—also a white, middle-aged man—pretends to be the originator of the website at an event in Amsterdam. Two years later, an advertisement for a performance at Postmasters Gallery in New York City claims that "the artist who created mouchette.org will not only reveal his/her/their identity and artistic motives, he/she/they will donate the web site to another artist who will take on authorship and continue the project."[102] As digital artists MTAA portray the event, "There was a big inflatable environment which contained Mouchette and people cued up for a chance to go inside and talk with (what turned out to be) him. He was a very polite, fairly young frenchman who took our picture after we talked for a few minutes."[103] However, another website linked to Mouchette.org counters this identification with "I think I understand your motives for pretending to be a 'middle aged white man' . . . or for that matter a 13 year old girl from france living in Amsterdam . . . than who you really are . . . A 13 year old black girl from Haiti living in an unglamorous neighbourhood in a metropolis by a polluted great lake; Toronto."[104]

While individual contributors may insist that Mouchette is really this or that person, the entire project offers contradictory explanations for who and where Mouchette is. The collective declaration that Mouchette is *a* and *b* and *c* is not much different from Jolles-Warburg's flowery confession that he sees the nymph *Pathosformel* here and there and then somewhere else. In fact, users access Mouchette's identities as a series, even though the site presents them as a network, since clicking on links produces a path (one of many) through Mouchette's personas. More precisely, participants experience Mouchette as a series of series.

Mouchette's seriality isn't disrupted by Neddam's convincing case that she is the true originator of the character. The artist continues to add new Mouchettes; she writes about a recent sighting of the character as a bas-relief of a little girl in Tbilisi, Georgia (figure 3.2). The stone carving was anonymously marked by painted graffiti of tears, a heart, and although Neddam doesn't mention it, two white blotches on the girl's chest and another where her crotch would be. "These graffitis [*sic*] make her an even a [*sic*] more suitable Mouchette, having travelled from one medium to another like mine did from film . . . to website, from high art to popular expression, from recognized art to participative, anonymous, humorous expressivity."[105] An image in transit, Mouchette's liveliness emerges not only from the openness of any individual incarnation, but also from the endlessness of her being and the ease with which she can be exported, reformatted, and shared.

The threesome of Warburg, Jolles, and their nymph becomes, in the case of Mouchette.org and its linked sites, an incalculable number of participants

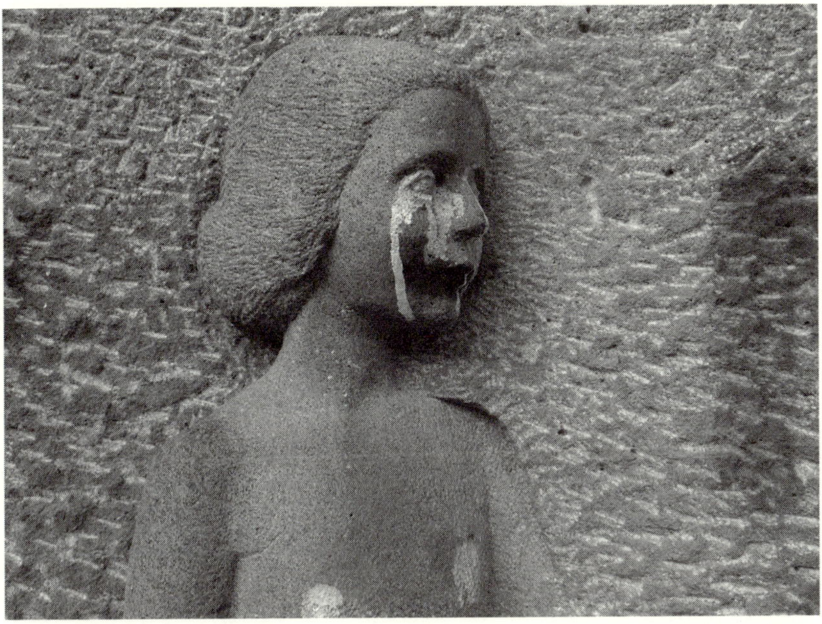

FIGURE 3.2 Martine Neddam, "Mouchette in Tbilissi" [*sic*], 2011. Courtesy of the artist

in dialogue with Mouchette and, importantly, each other. Per the latter, contributors to the suicide forum sometimes reference the comments of others. To give one example, on April 16, 2012, "mina aka seductress aka vampira" posts, "i miss you so much. i cant stop thinking about you. come back to me. please sweetness. the phone is off. write a letter." "Troy" responds 13 days later: "Mina, its Troy, readd me on msn/my new msn is sexygovernor@hotmail.com/or email me at my hotmail. missing You :P."[106]

The bottom of the forum page allows the viewer to read the comments arranged according to "famous users." After clicking on "Lucy Cortina," one of nine choices, I see a list of posts from 2002 to 2009, some written by her, others referencing her. In one of her earliest statements, s/he explains her motivation: "It's like this. I've decided that I'm going to make regular posts on this website. And to be honest, suicide is far too boring so I'm gonna try and cheer up you people, and make you realise that life is worth living. And I'm gonna put a title to my posts."[107] In one of the last ones, someone posting as "OMG, who's lucy cortina" demands, "LUCY CORTINA, I WANT TO TALK TO YOU."[108] The suicide forum in particular fulfills Neddam's desire "to create a social space" and "a platform of exchange."[109] Mouchette is variously described as a field of engagement, a "container" to be filled, something that travels by way of a human "carriers," and a "collective creation . . . built little by little through exchange and dialogue with visitors."[110]

The economy of producer to creation in Mouchette.org is many: 1 (one who transforms), a ratio that Teri Silvio considers to be characteristic of the paradigm of animation, more generally. Silvio's "Animation: The New Performance?" presents the former as a structuring "trope for human action on/in the world," essential for understanding twenty-first-century subjectification.[111] Contemporary media formations — such as the move away from auteurist cinema, rise of digital media, and increasing acceptance of the role of fans in the collective process of vivifying characters — have altered the ways in which subjects relate to each other and their mediascapes. Fueled by the shared ecstasy of fandom, Jolles and Warburg's collaboration is a small analog prototype for the large-scale collective animation of networked, socially driven plastic images.

vi.

Deep at the heart of the animated spectacle is one firm truth — behind every great animation, is an algorithm that is constantly being rewritten, a dream being reworked and a world being unmade. — Judith Halberstam, "Queerreality"[112]

One of the pages on Mouchette.org is entitled "Birthday."[113] There Mouchette quietly belies her own threats of suicide. As she explains, "My_Last _birthday_Party is a net.art group show which took place in Amsterdam in De Balie, on the 14th of December 2001, and will happen again in other places of the world in 2002, 2003, and forever. It's my last birthday party before I commit suicide. Everytime [*sic*] the last one again." Mouchette will always be on the verge of killing herself — she will always be an at-risk girl — but she will never do it. She celebrates the fact that every year will forever be her last.

That is, as long as someone continues to care for her. As Neddam admits in an interview from 2009, she still spends ten to fifteen hours a week on the site, even when she isn't creating new pages. "If nothing happened the art would die."[114] One of the benefits of Neddam's public admission that she is Mouchette.org's originator is that the effort it takes to maintain the ephemeral can be understood. Mouchette's perpetual tumbl(r)ing-toward-death requires a continuing investment in her vivacity. I am reminded of a mawkish but disconcerting conversation from *Rozen Maiden* (2004–6, 2013–ongoing), an anime series about living dolls with Victorian spiral curls and petticoats, heavily referencing *Alice's Adventures in Wonderland*:

> Doll maker: "As long as you pour your love into her, she'll go on living."
>
> Boy: "And if you don't?"
>
> Doll maker: "Well, if the love for her ever does vanish, she'll just go away, and she'll become lost, never to return."[115]

In the case of the Mouchette project, the loss of love isn't enough to banish the animated girl. Love *and* hate (and humor and empathy and profanity and absurdity and girlphobia and girlphilia) keep her alive. After all, van der Mersch's attempt to shut down the website only made Mouchette more vivacious. Bresson's widow addresses her cease-and-desist letter to the character and not Neddam, bringing Mouchette to life as a subject under the jurisdiction of the law. In other words, participation is what really matters, loving or not — that and Neddam's eighteen years of weekly labor maintaining domains, organizing comments on the forum, and so on. The Mouchette project is endurance performance art.

While the sites' flagrant resistance to the kind of planned obsolescence we accept in regard to digital media suggests a stubborn, even obsessive, duration, the youth of a girl character who is "nearly" a teen, her threats of just-about-goneness and embrace of the fly as her namesake emblem, as well as the broken links, "404 Not Founds," and "This webpage is not availables" emphasize brevity. In other words, the Mouchette project encourages a consideration of the "enduring ephemeral" that Wendy Hui Kyong Chun sees as essential to the ontology of the digital.[116] Challenging the idea that digital media hold information in storage (as we conventionally understand both holding and storing as keeping something still and intact), Chun reminds us of the active, "*time-based*" processes that allow computers to archive information.[117] "Digital media . . . constantly transmits and regenerates text and images."[118] It also loses things. It must. Indeed, the loss of loss causes problems. Right now, I'm typing on a computer that doesn't know when to let go. It holds onto information it should be deleting, allowing files to clog its hard drive and prevent proper functioning. I have to shift important files to an external hard drive or delete them in order to save this document I'm writing and avoid my Mac's dizzying, girlish spinning rainbow (aren't frozen programs surprisingly animated?). Periodically opening a Finder window just to watch my available gigabytes decrease, I am now always aware of the processual nature of computer memory and digital media.

Neddam's announcement may have spoiled the party for some who would rather believe that that the project was really originated by a child.[119] It also points to the many micro- and macro-movements that allow Mouchette to remain frozen at twelve years old, bringing the tensive union of the enduring and the ephemeral to the fore. In the article "Zen and the Art of Database Maintenance," Neddam works through, in the most practical and prosaic of terms, what would happen to the site if she were to die:

> My webhosting is paid by the year and is due for renewal in August. My domain registration is paid for two years and is due for renewal in February. The registrar will delete the domain name immediately after expiry but at least the webhost will tolerate one or two months of unpaid hosting before deleting the site. My credit card number can be renewed at least one more year without my intervention. My credit card is renewed every two years, in January. If I die now, how long will my site stay online and what will be removed first?[120]

For those who forget about the variability of websites, the Mouchette project can "stri[p] away the illusion of permanence and solidity, and expos[e] the frailty of a medium that relies on software instructions to create the appearance of a consistent environment" (to steal a quotation that originally referred to Mark Napier's Internet art piece *Shredder 1.0*).[121] However, Mouchette.org manages to dispel the illusion without completely discarding its promise. It vows to endure, but only through the transitory activities of participants, the inconsistency of an unstable art object, and the vulnerability of an at-risk — and, considering the fan shop, commodified — girl. In other words, unlike Tiqqun, who describe their mode of critique as a sexualized violation of the Young-Girl of capitalism, it refuses to strip the girl of her finery.

four. Exporting Girlishness

Otaku cannot resist the temptation to fiddle with fictional sexuality—reversing, recombining and otherwise reshaping it into endless variations. In this sense, the beautiful fighting girl is a remarkable invention of otaku bricolage. Her universality as an icon is a demonstrable fact. She has propagated herself all over the world through the Internet, and the seeds of a new generation are sprouting everywhere.
—Tamaki Saitō, *Beautiful Fighting Girl*[1]

The young girl in anime is always in the position of catalyzing some kind of emergent change.—Tamaki Saitō, *Beautiful Fighting Girl*[2]

Entertainment is Dead, Long Live Transformation
—Grant McCracken, *Transformations: Identity Construction in Contemporary Culture*[3]

i.

Despite—or, rather, because of—the proliferation of animation on the Internet (GIFs, flash animations, etc.) and at art galleries (animated video art), the movie house and the home theater remain remarkably generative sites for the consumption of plastic images: the expansion of animation outside its traditional environments assists the global dissemination of cinematic and televisual formats. With anime, the accessibility of contextual information—the vast amount of official and peer-to-peer expertise available on the Internet (not to mention the ease with which even obscure animation can be purchased through online commerce)—gives Japanese cartoons a secure foothold in markets outside Japan. An anime feature encountered on television can sometimes be rewatched, whole or in part, on YouTube, which might offer a sidebar list of related videos that includes anime-influenced video art and, more likely, Japanese television serials, in turn, motivating dedicated viewers (that is, otaku) to locate fan websites, attend screenings of Japanese exports at the local cineplex, and buy anime on Blu-ray from Amazon.com.

Some of these otaku make and share their own anime remixes, fan dubs, or fan fiction and thus become "multipliers," Grant McCracken's term for prosumers who participate in "collective acts of construction" and "use his or her instruments and networks to publicize" their efforts.[4] A small but active subsection of multipliers joins the Organization for Transformative Works, a group committed to protecting the rights of fans to create their own pieces using copyrighted characters or fantasy worlds. It takes its name from the legal distinction between a plagiarized and a derivative work: the latter is metamorphosis; the former, stealing.

For some this "transformational turn," as McCracken dubs it, has been an easy transition, one that seems to have happened the moment we turned our heads to program the VCR.[5] Yet questions nag. What is film after film after videocassette? What has been lost, gained, or willfully forgotten? What has been made vestigial? What marches on? What's at stake (or at risk?) when Manovich declares that "digital cinema is a particular case of animation" and McCracken proclaims that "entertainment is dead"?[6] Depending on who you are, these issues are film studies's anxious obsession, humdrum daily ritual (requisite morning tooth-brushing), or life's work — or at least, its most impactful contribution to discourses of mediation.

Animation is always present, explicitly or implicitly, in debates regarding the nature of cinema in the twenty-first century. More to my point, girlishness is an essential but unacknowledged element of the discussion, considering the adolescent qualities of the plastic image (as I have demonstrated throughout this book) as well as the sheer number of girls and other juvenilized creatures who have long been featured players in cartoons. While chapter 2 looks at girlish malleability vis-à-vis the prosumption and shareability of image files, and chapter 3 surveys the discourse regarding the girlish connectivity, vulnerability, and openness of websites and the Internet writ large, this chapter will engage with a set of "girl problems" that affect our understandings of what film is, where it travels, and — most importantly — what may or may not subsume it. These questions prove inextricable from our concern and desire for the Internet's ravenous consumption and metabolism of other media and the digital image's capacity for transformation. The animations I examine here put the girl at the center of such debates. Or, rather, they remind us that she was already at the center, but either too small, too micro, for us to notice, or too expansive, too obvious, to mention.

Instead of revisiting yet again the primal scene of English-language animation studies, the films of the Disney Studio (which I give their due in chapter 1), I'll start this chapter with an analysis of the historic anime feature *Ghost in the Shell* (1995), from which Huyghe and Parreno's *No Ghost Just a Shell* takes its title and little else. An adaptation of Masamune Shirow's cyberpunk manga, Mamoru Oshii's *Ghost in the Shell* is both an important work in its own right and a key player in historiographies of Western film; a global export, its aesthetic has inspired special-effects-driven blockbusters from all over the world as well as animations combining cel and CGI techniques. Although my book considers girlishness primarily from a Western perspective, this chapter acknowledges the difficulties of severing American and Japanese animation, in particular. Moreover, understandings of the shareability of girlish images are attached to conceptions of globalized consumption and of the Internet as a World Wide Web. Mouchette.org/.net, for example, is successful as a work of Internet art because of the global appeal of mutability, while at the same time, it attempts to produce mutability as a globally legible phenomenon (despite the linguistic specificity of the sites' primarily English- and French-language text). Likewise, the discourse promoting the Internet's inherent vulnerability is powered by anxieties regarding the transnational nature of the World Wide Web. And I believe that the ease with which Japanese and American animation have commingled has less to do with historical circumstances and innate similarities (although those factors do matter) than with a faith in the lightness, speed, and, above all, adaptability of plastic images. Any analysis of the plastic image's promise of flexibility, even one with an admittedly Western center of gravity, must address those images that have had a profound influence on the development of plastic media and of lability as a function and facilitator of globalization.

Let me skip right to the end of the film: *Ghost in the Shell* closes with an anomaly. Our cyborg heroine Major Motoko Kusanagi now occupies the body of a schoolgirl (figure 4.1). While this in itself isn't strange—as "ghosts" can be dubbed into any "shell" in this animated world—what is different is the child's voice that emanates from her body. Until now, the voice of the character has remained the same no matter what body she or he inhabited; this vocal consistency reinforces the dogged mind-body dualism of a film about the mobility of consciousness. Indeed, in an earlier scene, the hacker known as the Puppet Master, a sentient computer program gendered

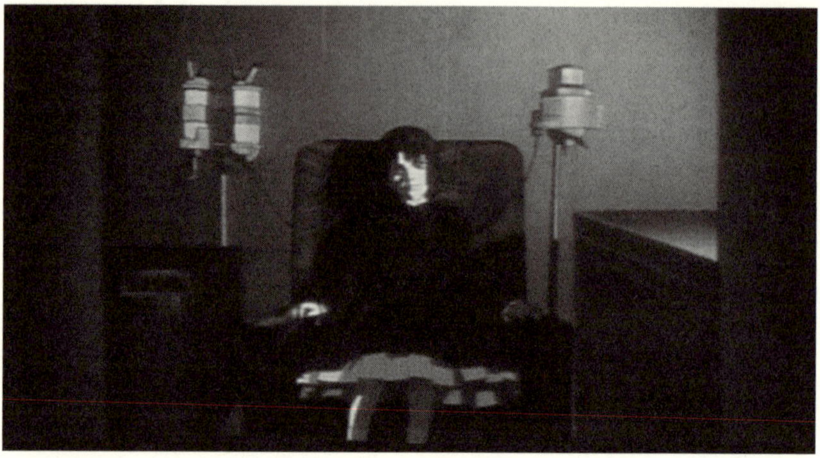

FIGURE 4.1 Mamoru Oshii, *Ghost in the Shell* (still), 1996

as male, takes the form of an aggressively beautiful gynoid, while his voice retains its detached and decidedly masculine timbre (in other words, paralinguistic markers are essential to character identification in this film). At the end, though, something weird happens: we both see and hear Motoko as a girl.

A prior scene provides some context. Before taking residence in the adolescent form bought off the black market, she appropriates the shell occupied by the Puppet Master, who retaliates by taking control of her now empty young woman's body (in a wry perversion of switched identity comedies). Then they engage in a bewildering conversation about the Puppet Master's scheme to merge their ghosts. He wants to mate with the Major, he explains, so that she can "bear [their] various offspring into the Net"—"a reference not only to cyberspace," Susan Napier clarifies, "but to a kind of non-material Overmind."[7] In other words, a masculinized computer program with a female body asks the gynoid currently residing in his former (female) body to give birth to their techno-progeny on a global information network. While we don't hear Motoko agree and don't know if she even has a choice, the next scene shows us what may be the consequence of their union: a doll-faced little girl in a black dress and white petticoat. The literal explanation of her new appearance is that her former bodies are annihilated by gunfire and her partner, Batou, can find only a substandard product on such short notice.

However, the Puppet Master's grand plan to use Motoko for some kind of cyborgian reproduction suggests that her resultant child form has greater significance.

The consonance between girlish body and girlish speech doesn't last long. In less than a minute, the child's voice disappears, modulating back to that of Motoko. She tells Batou that she is "neither the woman who was known as the Major nor the program called the Puppet Master." She soon wanders outside, gazing at a twinkling vista of the city. We learn in *Ghost in the Shell 2: Innocence* (2004) that Major Kusanagi 2.0 relinquishes this girl-body and disappears to live inside the Net, only to return when she is most needed.

Napier reads the juvenilization of Major Kusanagi as an expression of "her loss of power and identity."[8] She offers two possible interpretations: either the film draws attention to the disempowerment of women by "emphasiz[ing] Kusanagi's dependent feminine status" after her marriage to (or is it a rape by?) the Puppet Master, or the film underscores "the vulnerability of all human beings."[9] Either way, the body and voice of the girl are stand-ins for helplessness and lack (an attitude we have already seen in responses to Annlee in *No Ghost Just a Shell*). For Napier, this girl is an evolutionary step down from the capable and vigorous adult Motoko was before.

It seems to me, though, that while the end of the film does give us a startling vision of her new form — small and, at first, limp and fragile — it also introduces the schoolgirl as the final step before her trans mundane identity on the global Net, which in the film is a "transcendent realm of limitless possibilities."[10] This Net can't be visualized directly. The green scrolling numbers and other images on computer screens throughout the film are indices and not icons of the techno-spiritual world; they can only point toward a cyberspace that is "ultimately beyond representation."[11] With this in mind, Motoko's schoolgirl body and tiny voice can be considered visual and aural indices of a metamorphosis that can't be depicted literally.

My interpretation puts this scene in line with Napier's subsequent observations regarding the status of the adolescent girl in Japanese culture (an assessment that she does not bring to bear on her analysis of *Ghost in the Shell*): "In contemporary Japanese society, girls, with their seemingly still-amorphous identities, seem to embody the potential for unfettered change and excitement that is far less available to Japanese males, who are caught in the network of demanding workforce responsibilities."[12] Likewise, Sharalyn

Orbaugh, John Whittier Treat, and other scholars position the adolescent girl, the *shōjo*, within a space that is "liminal, transformative, liberatory, and potentially resistant."[13] As Orbaugh summarizes, while some Japanese commentators bewail the perceived girlishness of consumerist Japan (in an argument analogous to that of McLuhan's *The Mechanical Bride*, I would add), others embrace the *shōjo*'s "innate power to transform into other ontological possibilities, unaided by any outside force."[14] In a related discussion, Yuko Hasegawa considers the power of *kawaisa*, or cuteness, to indicate a "third sex" with the "potential for transformation."[15] Motoko's metamorphosis into a *shōjo* prefigures the "unfettered change" she will soon experience when she disappears into "the vast and infinite" network. While the Japanese *shōjo* and the Western adolescent girl are not exactly symmetrical concepts, the capacity of each to embody transformation and potential, however these ideas are understood in their respective contexts, is exactly what allows them to be evacuated of their more nationally and culturally specific meanings.[16]

Let's see what emerges from thinking girlishness as connectivity and not lack. What if the girl is not an image of Motoko's degradation, but both a prophecy and an embodiment of her development into data? What if she were the fragile shell and high-pitched voice that galvanizes Motoko's ascent to a networked plane of existence? To repeat something from chapter 1, what does the telescope of the girl allow us to see, especially in a world in which the most spectacular transformations can't be directly observed?

This brief but critical scene begins with a sluggish awakening. The image on our screen blips and glitches for a moment, as if our eyes are a malfunctioning video camera. Something seems to be there, but we can't quite make it out. The static quickly subsides, though, and we discover that we are looking across a room and into the next at Motoko-as-girl slumped over and attached to medical equipment. She lacks the mopey cuteness, shiny eyes, and bright coloring of *No Ghost Just a Shell*'s Annlee, but they are of the same genus. We soon understand that she just woke up and that the momentary video static is the technological equivalent of the bewilderment and hazy vision we experience after emerging from deep REM sleep. However, how can we look both through and at her eyes? How can Motoko now be both subject and object? As the "camera" advances on the unmoving girl, slowly, like a mechanical tiger stalking its prey, we realize that it's showing us a reflection of Motoko in a mirror. The effect is discomfiting, leaving us uncertain as to whose perspective is being offered to us.

Although she focuses on femininity and not girlishness, Japanese studies scholar Livia Monnet can help us work through the implications of Motoko's disorienting union with the Puppet Master. Monnet understands the fusion of these two cyber-minds in relation to the dissolution of boundaries that characterizes Sergei Eisenstein's concept of ecstasy (which, as I discuss at length in chapter 1, is an idea intimately entwined with girlishness) and Luce Irigaray's theory of the fetus-mother relationship. In this way, Monnet intervenes in the field of intermediality studies, which gives little attention to issues of gender or other modes of identification. Additionally, Monnet proposes that "the merger also embodies, and functions as a sign for, the potential of animation — in particular of digital animation — to transform existing realities, or reality effects, and create new worlds with which the spectator can identify in an ecstatic experience of the sublime."[17] Here animation is both a medium with its own unique capacity for plasticity and an intermedial, implicitly feminist practice "that has made possible the constant interaction or convergence between, within and across various media in the twentieth century."[18] The merger thus has multiple, interrelated meanings in Monnet's analysis: most literally, the union between the Major and the Puppet Master, but also the ecstatic bond between mother and fetus, the blending of two images to produce a digital morph, *Ghost in the Shell*'s combination of CGI and cel animation, and the integration of analog and digital cinema in *The Matrix* (1999), inspired by *Ghost in the Shell* and *Alice's Adventures in Wonderland*. Most evocatively, the merger refers to the production of an intermedium that is more than the sum of its parts.

For Monnet, *Ghost in the Shell* also "suggests that digital animation, like most recent new media art, computer games and other new media objects, requires new conceptions and modes of presentation of fantasy, subjectivity, corporeality and the self."[19] If so, then perhaps the girl's small, limp body and border-crossing visual perspective — her simultaneous but dispersed vision of self and other — are part of this new fantasy of identity (or, rather, a newish fantasy of identity, as I argue throughout this book for the close relationship between digital and protodigital forms). Perhaps girlishness is hybrid animation's essential "mode of presentation."

The final scene of *Ghost in the Shell* is an anticipation of the centrality of girlishness to its sequel. However, the girl to come is less an icon of the limitless possibility of the Net than a symbol of globalization's exploitation of such possibility, bringing the long history of the girled commodity to the fore. The plot of *Innocence* involves the kidnapping of little girls, stealing of their ghosts, and uploading of these psyches into expensive sex dolls sold by a multinational corporation. A clever hostage makes the gynoids turn on their owners and then self-destruct, resulting in a spree of spectacular murders. *Innocence* shows us what happens when girls become chattel and, later, weapons through a transnational network of exchange: the girled product strikes back, unleashing some of the force contained by the little cyborg at the end of the film's precursor. However, while we sympathize with the kidnapped girl who engineers the killings, we are asked to see her lack of remorse as problematic. Unlike Huyghe and Parreno's *No Ghost Just a Shell*, *Innocence* suggests that perhaps the girled product can be saved and even reformatted, but not reformed.

The end of *Ghost in the Shell* is also a fleeting revelation of the girlishness that powers another animation, Satoshi Kon's *Paprika* (2006), which was also released globally and has received popular and critical acclaim.[20] Indeed, the passive girl-body plopped on a chair in *Ghost in the Shell* holds a force that later explodes across the screen. While these are very different films — as William O. Gardner reminds us, *Ghost in the Shell* presents the Net as an immaterial and transcendent elsewhere, while *Paprika* understands its own mediascape as always intervening in so-called real life — both deal with the pleasures and perils of the Information Age. And both offer the girl as an image of the transformations facilitated by digitality. *Ghost in the Shell* gives us a momentary, anomalous vision of the girl's potential energy, *Innocence* warns us against the squandering of that immense potential, and *Paprika* is that potential turned kinetic.

This chapter uses my reading of the technologically evolved girl at the end of *Ghost in the Shell* as a launching point for an analysis of the girlishness of *Paprika*. Or, to be more precise, Oshii's now classic example of dystopian, "woman-centred" cyberpunk provides gravity assistance for Kon's later exploration of mind-hacking and plastic identity.[21] Gravity assistance is the process by which spacecraft can gain (or, less frequently, lose) velocity by using the gravitational pull of a planet. *Ghost in the Shell* is a celestial body that produces a powerful gravitational pull. When the figure of the girl in *Paprika* passes by this body as she circulates both transnationally and transmedially, she picks up speed. The eponymous Paprika whips by Motoko-as-girl and accelerates into a new orbit. To take this metaphor even further: in certain ways, Annlee from *No Ghost Just a Shell* is the force of gravity that connects a planet to its satellite moons (or the work of the originating artists to that of their collaborators). Mouchette from Mouchette.org/.net is a universe of habitable planets. Paprika, on the other hand, is the rocket that travels throughout the media solar system.

Kon's animation tells the story of a psychologist who uses an alter ego named Paprika to enter her patients' nightmares. This therapeutic method requires a special machine, the DC mini, and a computer equipped with software that mimics the interface of nonlinear video editing programs like Final Cut Pro. The playful, emotional, and more youthful foil for the exhaustingly sober Dr. Chiba, Paprika is a superheroine whose power lies in her ability to modulate according to the logic of dreams. Importantly, the dreams depicted in the animation incorporate various media formats and viewing practices within the diegesis — a Super 8 film projected on a wall, famous Western feature films exhibited in theaters, movie posters, websites, and billboards — all visualized through *Paprika*'s metamedium, a combination of cel and 3-D CGI animation (often called 2.5-D).

Paprika is the only character who can effectively navigate these complex mediascapes. She can scale herself to fit inside advertisements. She can leap in and out of the films-within-the-film, metamorphosing into characters appropriate to the particular movie's mise-en-scène. Sometimes she resides inside the space of a website that looks like a cocktail lounge and is called a radio club. And at *Paprika*'s end, she transforms into a giant baby and literally consumes her foe. Devouring the evil Chairman, she rapidly develops from infant to adolescent and then disappears (figures 4.2 and 4.3).

Paprika proffers a visual theory of media convergence as not only an issue of technology, but also one of globalization. When the Chairman steals the DC mini and uses it in both directions — to enter people's nightmares and to let their nightmares escape into the real world — the dream imagery forms a riotous parade that runs rampant through city streets. This ever-growing procession comprises appliances, furniture, Victorian bisque dolls, Lady Liberty, a Venus de Milo sculpture, statues of the Virgin Mary and Buddha, Russian matryoshka dolls, and other humanoid toys and figurines. Therefore, *Paprika*'s vision of media convergence is one in which boundaries between cultures, technologies, commodities, and people are horrifyingly permeable. While our supergirl is eventually able to stop the parade, first by turning brand animation against itself and later by consuming the Chairman, these multiple transgressions cause mass confusion, madness, injury, and death. As Thomas Lamarre says of Hayao Miyazaki's animations, "Only a girl can save us now"— or, rather, only a girled image can save us.[22]

While the film is perhaps too simple in its judgment of media convergence, it is complex in its investigation of the intersecting processes by which media can come together and of the role of the plastic image in consolidating and even producing such amalgamations. As with *No Ghost Just a Shell*, Mouchette.org/.net, and, in Monnet's analysis, *Ghost in the Shell* and *The Matrix*, *Paprika* relies on animation's power to produce plastic images and gather media (although the discourse surrounding Huyghe and Parreno's work tends to focus on the gathering of a community of different kinds of artists and not the media they employ). All of these works also engage with issues of femininity and the digital image. However, *Paprika* is unique in regard to the thoroughness with which it explores digital media's imitation of older media, including contemporary branding's redux of the aesthetic of cel animation, through the plastic girl's special abilities. The film is persistent in its rhetoric: our contemporary condition is generated and maintained by a triangulation of digital media, analog media, and girlishness. I've used "our" quite consciously. In *Paprika*, the media mash-up, what I earlier call Girl Talk, is a transindividual and transnational conversation — indeed, anxiously so.

ii.

Put simply, we are increasingly discovering that we and the world swim in a sea of media, and vice versa. And we are increasingly finding that the nature of not only what swims in that sea but of that sea itself is animation "as such," which for us it has always been.—Alan Cholodenko, *The Illusion of Life II*[23]

We can no longer place our dreams in boys. People are seeking catharsis in the fighting of the weak. — Toshio Okada, quoted in *Beautiful Fighting Girl*[24]

Paprika is loosely based on a science fiction novel by Yasutaka Tsutsui, which was first published in 1991 as a serial narrative in the now defunct Japanese version of *Marie Claire*. It is fitting for my purposes that it appears in a globally franchised fashion magazine for women and that Kon's film adaptation did not premier in Japan. In fact, the animation's first screening was at the Venice International Film Festival, an important stop on the festival circuit, where it was nominated for a Golden Lion. *Paprika* is very conscious of its place within the global image economy. Like much contemporary anime, it was produced with both local and global consumption in mind. Susan Napier explains that "despite its indisputably Japanese origins, anime increasingly exists at a nexus point in global culture; this position allows it to inhabit an amorphous new media territory that crosses and even intermingles national boundaries."[25] Hiroki Azuma advances a controversial argument that Japan's culture of animation was instead imported from the United States after World War II; now the global trajectory of animation has reversed and multiplied, facilitated by the worldwide "postmodernization of culture that began in the middle of the twentieth century."[26]

Susan Pointon uses the phenomenon of anime to challenge the assumption that cultural imperialism applies only to the unidirectional movement of products from America to the rest of the world. Her analysis is less a critique of Japanese production and marketing practices than an acknowledgment of the complexity of globalization: "From a theoretical point of view, the study of a media phenomenon like anime points up the difficulty of containing contemporary cultural texts within strict national borders. It is impossible to ignore the constant cross-pollination and popular cultural borrowings that complicate and enrich the anime texts."[27] And Steven T. Brown, also underscoring the connections between the rise of anime and broader sociocultural transformations, goes so far as to claim that

> if one considers the incredible rate of consumption of anime outside of Japan, in regions as diverse as East and Southeast Asia, North America, Europe, and Australia . . . one would have to say, in an era of global capitalism and the accelerating transnational traffic of cultural flows, that *the world is the anime screen*.[28]

Some images flow around the world faster than others. As I have argued throughout this book, the assignation of the qualities of availability, shareability, and lightness to children and, especially, girls readies their images for transport across national and cultural boundaries. Look at a representative issue of the anime, manga, and gaming magazine *Otaku America*: pictures of children — feminine adolescents, mostly — as well as animals, monsters, and cyborgs dominate its pages. These unfinished beings and nonhuman creatures are doubly girled through animation's (and manga's) connectivity and openness, or what Marshall McLuhan calls its coolness. In his schema, cool media like cartoons solicit an active mode of spectatorship by refusing to provide high-definition sensory information. Through a problematic but nonetheless instructive analogy I discuss in chapter 2, McLuhan compares cool media to girls wearing dark sunglasses, which produce the void "that invites a great deal of participation and completion." Men are attracted to girls in shades, he claims, but "seldom make passes at girls who wear [prescription] glasses."[29] If the cartoon is an image that gains its power through the dark emptiness of its metaphorical orbits, then anime, with its often abstract draftsmanship and purposely simplistic rendering of movement, takes full advantage of this lack. This isn't to say that there aren't many animations featuring adults or those that counteract the unfinished, immature nature of the cartoon image by imitating cinema. Instead, I propose that mature, autonomous images don't explode across the world screen with the same force and acceleration; they appear not to want, and, most importantly, not to want *us*. The unfinished image asks us to help it along, and when we oblige, it picks up speed.

Paprika's attractiveness is largely due to the motivity of its eponymous heroine. Indeed, the film is at its most engaging when Paprika flies across the screen (she "runs away with the film," write reviewers Timothy Perper and Martha Cornog).[30] The opening credits emphasize mobility, flexibility, and scalability from the get-go, establishing the promise of plastic images before we are allowed to understand their threat. A detailed description of this early sequence will demonstrate the particular capacities of the girl that come to good use as the film proceeds. Especially of note is the scene's coupling of kineticism and transformation, suggesting that the plastic girl's perpetually unfinished form — her coolness — is related to the nimbleness of her body.

In the opening credits, Paprika takes us with her on a rip-roaring journey through a city at night. She starts out on a red moped. Looking to her right, she sees a painting on the side of a commercial truck: a cutesy little girl in pigtails straddling a rocket that bears a passing resemblance to the animated shark Jabberjaw. Suddenly the girl on the rocket comes to life: we realize that she is now Paprika. After the projectile rears back on its tailfin, they both zoom off the side of the truck and over the top of high-rise buildings, passing two giant billboards. Now Paprika appears inside a cigarette advertisement on the left, and, in a moment, pops up as a beer drinker in the advertisement on the right (plate 20). Suitably, she changes outfits with each role. She then moves out of the frame of the billboard and into a computer monitor on the desk of a sleeping man slumped over in his cubicle. We get a glimpse of the open windows on the desktop of his computer. It is late for him to be working. Stepping out of the screen, she places his jacket back on his shoulders. She is wearing a white nightgown, fitting for such a sweet gesture, but will soon sport her usual red shirt and jeans.

Her childlike joie de vivre incompatible with the solemnity of the workplace, Paprika skips out of the office, becomes semitransparent (perhaps to avoid a security guard), and then regains her substance in order to cross the street. The city is hectic, dangerous. Nobody is stopping. So she snaps her fingers to arrest the cars and pedestrians whizzing by and gambols across. Now she is sinking her teeth into a fast food burger; two cocky teens try to flirt with her, but she dodges them by jumping into the decal of the T-shirt of a boy skating on the sidewalk. After briefly reclaiming her moped, which definitely suits her, she is pictured driving a car. There's a cut. The driver is now Dr. Chiba. The girl's open, curious face is replaced by the weary eyes and down-turned mouth of her very serious alter ego.

This sequence shows us the capacities that allow Paprika to traverse the treacherous mediascapes later in the film. First, she moves often and fast (although she is usually dependent on some kind of vehicle). Second, she can enter any image and scale herself accordingly. To use an analogy from computing, she can export herself in different formats appropriate to various software applications. This permits her — and often requires her — to play roles pertinent to the image at hand (to drink beer in a beer advertisement, for example). She is, like Huyghe and Parreno's Annlee, a "polyphonic character," and her ability to blend into any image later proves to be an effective form of camouflage.[31] Similarly, she can move between and into different dimensions — that is, from the (simulated) three-dimensional space of the city to the two-dimensional space of the billboard and the side of the truck.

Her multidimensionality provides an additional service: the visual cacophony of late-modern urban space becomes quieter. Instead of a raucously disjunctive visual field — of billboards on top of buildings in front of other buildings — we have a series of related images. Her presence within all of them connects one to another as continuity editing creates smooth connections out of film cuts. We are reminded of the advantage of this particular skill later in the film when one of her patients has a panic attack while stuck in traffic. His eyes dart back and forth between the commotion of the city during rush hour and various advertisements on buses and buildings. His visual field is anxiously fragmented, his mental state is disintegrating, and Paprika isn't there to put him back together.

Paprika is an animated character who is of a piece with the medium of animation, itself, which can offer continuity between different formats (2-D cel, 3-D CGI) and styles (graphic, more volumetrically realistic, etc.). The obviously constructed nature of animation makes almost any particular rendering as believable as the next. While this can take the form of *Paprika*'s 2-D cel animated characters and 3-D CGI environments, it can also be more spectacular. The benchmark television anime series *Neon Genesis Evangelion* (1994–95), for example, features very disparate animation styles that somehow hold together. Sometimes shadows are sharply bounded shapes of uniform grayness; other times they are gradient images. Characters transform into rough, jagged sketches for a comic scene and later become frozen black silhouettes against a painterly background. In one episode, a silhouette of a woman in profile has a bright oval hole for an eye, as if white light in the background were shining through a tunnel bored through her head, as if McLuhan's dark sunglasses flipped up to reveal a cooler and more inviting void. The character with which she converses is also shown as a black shape in profile, but has no such hole.[32] It's really the ongoing presence of the characters, however their stylistic depiction, that allows this piebald world to cohere. *Neon Genesis Evangelion* establishes a logic within the animation that imitates and supports the rhetoric of the larger culture of anime consumption: "In the multimedia environment of the 1990s," and even more so today, "it is only characters that unite various works and products."[33]

Of course, animation need not be deployed in this fashion. Walt Disney is known for enforcing a uniform style among his artists. Kon distinguishes the look of American animation from Japanese anime in regard to a difference in divisions of labor; Japanese directors divvy up the film for the lead animators scene by scene, resulting in a consistency of style within scenes but not always between them, while American directors give each lead animator a particular character to ensure that the figure is drawn the same way throughout.[34] Here Kon implies that anime characters are imminent, not transcendent — they emerge from their immediate environments (the circumstances depicted within the scene). While this is a remarkable insight, he doesn't acknowledge the distinction between popular and art house American animation. Kon's assessment does not necessarily apply to experimental and avant-garde works, which tolerate discontinuity and usually have a different architecture of labor. While mainstream American animation does tend toward constancy, animators can choose instead to push their medium both far and in multiple directions at once without the animated film's complete disintegration.

Paprika is a cel-animated plastic image that functions as a shape-shifting intermediary between the different media presented within the film. When she moves from the photographic billboard to the computer screen, for example, she performs animation's ability to negotiate an exchange between our conceptions of analog film and digital media. Historically, cel animation has been employed to produce images of high-tech GUIs in science fiction films made before computer animation was developed or made financially viable. The computer screens within the film *2001: A Space Odyssey* (1968), for example, were created using cel animation. Therefore, photographically based cinema could often approach the digital image only through animation. Cel and other analog practices of animation also mediate between the concepts of analog and digital film in discussions of the status of cinema in the Information Age. When scholars such as Manovich position digital cinema within a lineage of cartoons, they implicitly render cel animation as the go-between allowing photographically based live-action film and digital cinema to speak to each other while still acknowledging their differences. Animation is the mediation of mediation.

Thomas Lamarre's discussion of the girl-robot Tima in Rintarō's *Metropolis* (2001) is instructive here. Indeed, Tima and Paprika perform consonant roles in their respective films. Registering the "tension between digital animation and cel animation," little Tima is a sophisticated android constructed to replicate the deceased daughter of a city's would-be overlord in an animation based on manga by Osamu Tezuka (1949) and Fritz Lang's film of the same name (1927).[35] Lamarre contends that she is a function of irreconcilable divisions and multiple origins. In fact, *Metropolis* amplifies her fraught identity and "lingers on the impossibility of fully determining the robot, who remains full of potentiality yet decidedly underdetermined."[36] This is a graceful description of Paprika's condition, as well. But Lamarre and I part ways when he chooses to bracket issues of gender, missing an opportunity to fortify his reading of *Metropolis*. Tima's potentiality and underdetermination can't be separated from her gender or, I would add, age. Her girlishness contributes to her success as the film's primary rhetorical figure in its consideration of digital-analog relations. Connective, vacuous, underdetermined, underdeveloped — literally underdeveloped, as Tima is not only a child, but one prematurely electrified into life before her capacities have stabilized — Tima is a perpetually unfinished emblem of new media's ongoing attempts to clone certain aspects of analog film and a confident proclamation of the digital object's novel superpowers.

Though she dies near the end of the film, we later hear her vulnerable, breathy voice emanating from a radio: "Who am I?" she asks, continuing to question the ontologies of media even after her apparent demise. Her disappearance from view occurs after she loses her beatific demeanor and becomes demonic. Unleashing an "entire world-wide network of forces" upon the city before succumbing to destruction, she — or at least, her ghost — remains inside this network to implore the denizens of Metropolis as well as the viewer through a broadcast medium.[37]

Metropolis, Lamarre explains, is an animation about media, but it is not ultimately an animation about mediation. He sees mediation as a resolution between antinomies that also reinscribes their fundamental opposition. Here as well, our thinking diverges. I'm suggesting that mediation can be a kind of never-ending negotiation between parties that are both joined and held apart — a mediation without adjudication, we could say, easily applicable to Tima.

Paprika embodies this particular notion of mediation through her remarkable ability to produce continuity without effacing the lines between media. In other words, Paprika lets "differences between media remain visible, palpable, and sensible," to steal Lamarre's characterization of *Metropolis*.[38] *Paprika*'s opening credits make us acutely aware of various frames — the rectilinear contours of the billboards, the truck, the computer screen, the open windows on the man's computer desktop, the photographs push-pinned to his cubicle, the picture printed on the boy's T-shirt, and so on. Paprika can enter these spaces, but she can't obscure their boundaries; indeed, when she exits the computer, the chassis of the monitor — the plastic frame around the screen — cuts her head in two (plate 21).

The thick line temporarily bisecting her forehead underscores the applicability of what's called remediation to this scene and the film, more generally. Defined by Richard Grusin and Jay David Bolter as "the formal logic by which new media refashion prior media forms" and vice versa, remediation makes its mark on the credit sequence, which incorporates aspects of photography (the billboards), painting (the rocket girl on the side of the truck), and — to point to the remediation of newer media forms by older ones — the GUI (the computer).[39] This scene performs a mode of remediation in which media "are presented in a space whose discontinuities, like those of collage and photomontage, are clearly visible" (remediation as, in Lamarre's terminology, nonmediation).[40] *Paprika* wants to have it both ways, or, rather, it is about how *we* want it both ways: the frames morcellate the image, while the translation of every medium into animation keeps it together. Paprika's trajectory through the city links spaces of different dimensions, while their boundaries temporarily slash her in two. Paprika is always herself, but she is always changing.

The explicitness of this mode of remediation encourages us to reflect on the scene immediately prior, which depicts a recurring nightmare of one of Dr. Chiba's patients, police detective Toshimi Konakawa. Fitting for an animation that is centrally concerned with relations among media, his dream refashions scenes from several famous live-action films. It has a different way of emphasizing the media's enframing of images, however. There is no line that cuts across Paprika's frontal lobe. Instead, the frame is now the film cut that separates one clip (and one cinematic allusion) from another in the rapid montage. Both Konakawa and Paprika provide some continuity across the divide. However, it's the latter who can be at home here (or in all the "heres"). Konakawa experiences the dream as an uncanny force of psychic disintegration. Only Paprika can handle the constant modulation required to mediate mediation.

Konakawa's dream begins at a circus, in homage to *The Greatest Show on*

Earth (1952). He suddenly falls through the floor and is caught by Paprika, wearing a spangled pink costume and swinging toward him (and us) on a trapeze. Then they are pictured as Tarzan and Jane holding a vine and sailing through the air, only to tumble again and suddenly appear on a dark train from the James Bond film *From Russia with Love* (1963). Konakawa grunts as his enemy chokes him with a wire, but Paprika saves her patient by slamming the aggressor over his head with a suitcase. However, this movement sequence is cut short; her whacking motion is picked up in the next clip, in which her blow is completed when she bashes another man with a guitar. In the lingo of film editing, this is called cutting on motion.

Konakawa then takes a picture of Paprika and a crowd of people. They freeze completely when the flash goes off, as if actually transformed into a photograph. But someone manages to escape. Konakawa chases after him, only to end up at the end of a long hall, perhaps in a hotel. We hear a gunshot. The man falls forward in chilling slow motion. Konakawa lurches toward him, but the floor softens and starts rolling like the waves in a waterpark fun pool (a motif repeated in various dreams in the film). He screams as he plummets into bright white light, such as what we might see when viewing the tail leader of a film, and we hear the soft whirr of a film projector. "But what about the rest of it?" he calls out in voice-over, wondering why the dream ends without resolution. When the nightmare recurs later in the film, the "camera" pulls back at this moment to reveal a screen in a movie theater. Paprika, sitting alone in the audience, applauds while Konakawa falls through white nothingness.

He finally wakes up, panting and sweating in terror, and he and Paprika review his dream together using a software application with an interface that looks like that of a video editing program. Konakawa calls this combination of the DC mini (an elegantly curved accessory with spikes worn above the ear of the doctor and patient), the laptop, and the software "a perfect dream machine." Thus, in *Paprika*, the computer is a machine for viewing dreams, and dreams — what we can consider to be the oldest of old media — remediate photography and film. And though I am at risk of plunging us into the same hopelessly dizzying funhouse reflection that the film (and its successor *Inception*) attempts to visualize, Konakawa's dream also re-remediates that which early film appropriated to make a name for itself: train travel and the fairground attraction. *Paprika* uses a projection of future wearable computing to look back at the history of cinema through the means of the present.

iii.

The structure of the Internet reflects the postmodern world image. Therefore, the logic of the web, which is produced out of this structure, is in fact not limited to the world of the Internet and computing. As everyone knows, in the past few years the terminology of various cultural forms has suddenly slid like an avalanche, seeking outward appearances that seem Internet-y and computery. As a result, the logic of the Web is penetrating widely and deeply into many genres technologically unrelated to it. —Hiroki Azuma, *Otaku*[41]

Paprika fears the perilous fallout from merging, or sharing, dreams. Moreover, it is afraid of a loss of self in such a shared dream as the movies, or the Internet, or the modern world: to lose self in this world is to lose perspective of the world, and one's relation to it.— Ryland Walker Knight, "Return to the Movies, Return to the World"[42]

Paprika's oneiric recursion — dreams include photographs, films, and computers that are also machines for viewing dreams, photographs, and movies — becomes all the more vertiginous with the animation's explicit connection between dreams and the Internet. Just as the computational assemblage of the DC mini, laptop, and software allows patients to share their otherwise private dreams, the Internet, Paprika explains, is a form of dreaming made public (aptly, this dream girl hangs out in a bar that is also the website www.radioclub.jp). Here the film offers its most anxious criticism of twenty-first-century mediascapes. The proliferation of frames in his urban and oneiric environments triggers Konakawa's intense physio-psychological discomfort, but it's the mass sharing of dreams — that is, the Internet — that is truly disastrous.

The boundaries between media, which were compromised but still palpable in the exhilarating credit sequence, are dissolved as the film proceeds. We soon discover that *Paprika* not only agrees with Friedrich Kittler's claim that "the general digitization of channels and information erases the differences among individual media," but also adds that it erases the differences between individuals.[43] The film is concerned about both the centripetal and centrifugal forces of media convergence: the coming-together of different media through one platform (such as Facebook) and the spreading out of content across different platforms (what Nicholas Negroponte calls "multimedia," as I discuss in chapter 1). If "the age of media convergence enables communal, rather than individualistic, modes of reception," as Henry Jenkins insists, then *Paprika* warns us against joining the media commune.[44]

At this point, I want to draw a fine distinction between the frequently confused terms "remediation" and "media convergence," which focus on different aspects of the same basic idea. Jenkins, the scholar most commonly associated with the latter term, defines convergence as "the flow of content across multiple media platforms, the cooperation between multiple media industries, and the migratory behavior of media audiences who will go almost anywhere in search of the kinds of entertainment experiences they want."[45] As part of the same discussion, he looks to Ithiel de Sola Pool for an early but foundational understanding:

> A single physical means — be it wires, cables, or airwaves — may carry services that in the past were provided in separate ways. Conversely, a service that was provided in the past by any one medium — be it broadcasting, the press, or telephony — can now be provided in several different physical ways.[46]

Although Jenkins's theories tend to privilege the movement of consumers and content across media and not multimedia consolidation (the so-called black box school of thought), he admires de Sola Pool's recognition of the dual forces of convergence — the centripetal and centrifugal trajectories. On the other hand, Bolter and Grusin's remediation is the incorporation and refashioning of the codes of one medium by another.[47] The difference between the two is really a shift of weight: Jenkins is largely concerned with forces and choreography, while Bolter and Grusin are more interested in reconfiguration and recoding. However, movement does permit a medium to be remade by another, and, as Arild Fetveit writes, "a series of remediations allow established media like newspapers, photography, radio, film, and television to free themselves partly from the media technologies in which they were developed and migrate across any channel open to them."[48]

At its most perceptive, *Paprika* reminds us that the motion and morphing of media are inextricable. At its most alarmist, it neglects to account for the ways in which media objects stay put, despite the digital image's promise of plasticity and the larger cultural turn toward liquescence (the openness of channels is affected by planned obsolescence and access to hardware, software, and training, among other factors). But it's important to acknowledge that *Paprika* is less concerned with advancing media specificity than with illustrating the horrors of convergence, albeit through a metamedium that is itself a hybrid of the analog and the digital.

Paprika gives us an ugly image of the movement and metamorphosis of media. After the Chairman hijacks the DC mini to force open the portal between dreaming and waking, individual nightmares come together as a ramshackle and ever-growing parade of animated things: a refrigerator, a microwave, a boom box, bookshelves, Victorian dolls, geisha dolls, anatomical models, Japanese ball-jointed figures (the kind that influenced Hans Bellmer's photographs and Oshii's *Innocence*), a dinky Godzilla, a pink elephant, stuffed animals, dirigibles, a helicopter, a Virgin Mary statue, an elongated Buddha, an emaciated puppet, a gorilla similar to the game character Donkey Kong, pigs, the gate to a Shinto shrine, a chair, a suit of armor, frogs banging drums and tooting horns, and so on. Importantly, there are domestic products like appliances (autonomous things advancing upon us, divorced from their use), image products like the knockoff Donkey Kong, and things associated with cheaply commodified national culture (such as Russian matryoshka dolls; the Statue of Liberty, which might as well be a statuette sold on a street corner in New York City; and the Japanese Maneki-neko, usually a figurine or bank in the shape of a cat with a raised paw, a mainstay of global Chinatowns). These items look like overgrown plastic toys and giant versions of dollhouse furniture in the colorful aisles of a dollar store. *Paprika* uses this clownish procession of things as an analogy for the convergence of global media objects — that is, the "texts, sounds, and images [that] become detached from the media technologies that used to support them, and perhaps even gave birth to them . . . and migrate and multiply on technological platforms flattened by digital technology."[49] Not incidentally, this description of convergence begs to be visualized via animation or special-effects-driven live-action film; indeed, this sounds like a scenario for a film about film after film. Manovich uses equally animatic language: "It is as though different media are actively trying to reach towards each other, exchanging properties and letting each other borrow their unique features."[50] *Paprika* aptly claims that we understand processes of convergence and remediation as the animation of media objects and even the media, themselves. Animation is once again the mediation of mediation.

While some objects in the parade are their usual size — a chair and a coffee mug, for example — others are unnaturally small or large, distributing Alice in Wonderland's changeability across the group of figures. For example, a frog has grown to one-fifth the size of the Statue of Liberty, which has shrunk so its eyes meet a streetlight. They all march to the rhythm of aggressively upbeat circus music, the sonic equivalent of Stephen King's character Pennywise the Clown, and cut a shaky path through the country and later the city. The procession of varied and visually striking images has the appearance of order from far away. Once it gets closer, we see that it Tumblrs through space as what Seth Price might call a hoarding (just as surf clubs give us the neat chronologics of the series while also reveling in the supposed randomness of its images, a phenomenon I address in the previous chapter).[51]

While my description might conjure an image of fascist body culture, in actuality the choreography of the procession is too loose, too inebriated, as if a herd of rhythm-impaired zombies were forced to join the circus. This parade embodies the deleterious movement of dream imagery from one platform (sleep) to another (waking life) and from the private, protected spaces where Dr. Chiba/Paprika conducts therapy sessions to shared, public spaces (as the Internet has escaped our offices and homes to wander about via smartphones' wireless capabilities), as well as the merging of many into one (as the same smartphone allows us to watch videos, listen to music, check email, etc.). It also provides the visual and aural accompaniment to various acts of unwitting self-harm: trapped inside a hallucinatory mass dream, people jump out of windows, leap off buildings, and walk off balconies. "*Paprika* fears the perilous fallout from merging, or sharing, dreams," mapping the share function of networked media onto the share function of the DC mini.[52]

As you may remember, Monnet links merger to emergence, arguing that *Ghost in the Shell*'s sophisticated union of CGI and cel animation and its diegetic culmination in the blending of characters' ghosts over a global information network "embodies, and functions as a sign for, the potential of animation — in particular of digital animation — to transform existing realities, or reality effects, and create new worlds with which the spectator can identify in an ecstatic experience of the sublime."[53] In her analysis, the meta-medium of animation facilitated the convergence of media in the twentieth century (a process of mediation that gave rise to *Ghost in the Shell* and is depicted within its frame). *Paprika* uses its own synthesis of analog and digital animation and its diegetic presentation of convergence to produce another kind of ecstasy — Jean Baudrillard's "negative ecstasy," "an obscene delirium of communication."[54] Using the example of radio, Baudrillard complains that "the stations overlap and mix together (to the point that sometimes it no longer communicates at all)"; in fact, "the space is so saturated, the pressure so great, from all who want to make themselves heard."[55] Although "the pressure" in *Paprika* is the result not of a need to be heard but of the antagonist's desire to make others' private fantasies "overlap and mix together," Baudrillard's account of the role of "circuits and networks" in producing a world of grotesque explicitness, "availability," and "fluidity" accurately describes the particular form of apocalypse that the Chairman attempts to bring about.[56] Both Baudrillard and *Paprika* are deeply suspicious of the "narcissistic and protean era of connections, contact, contiguity, feedback, and generalized interface" (indeed, the babble of characters under control of the Chairman — a form of Girl Talk that seems, at first, to be bafflegab, but upon closer listen, is an accurate description of the dream imagery that at the moment only they can see — seems like a mordant rejoinder to criticisms of Baudrillard's writing style as raving postmodern gibberish).[57]

Dr. Chiba's colleague Chief Shima is caught in this collective nightmare. While his RL body runs through a closed window, his legs continuing to pump in midair, his dream self is fooled into thinking that negative ecstasy is delirious pleasure. He ends up a lolling master of ceremonies carried aloft by a giant mound of dolls. Paprika intercepts the parade to wake him up. However, Himuro, one of the creators of the DC mini and the first to fall under the Chairman's nefarious influence, thwarts her efforts. He keeps popping up as different dolls in the pile, attempting to convince her that Chief Shima is hopelessly lost. During this scene, Paprika wears a large bonnet tied under her chin and a frilly skirt. She's dressed in what's called Lolita fashion, similar to what most of the dolls are wearing (one of the best-known purveyors of the Lolita look is the Japanese brand Metamorphose Temps de

Fille, mistranslated on its website as "the time of transformation into a little girl." "Almost everyone has the desire to 'transform' one's self," the company avers.)[58]

Chief Shima awakens and identifies Paprika as the "beautiful little feisty peasant girl I saved sometime ago." She now has a new hairdo and outfit. Coyly, she promises "to show him her gratitude," crawls toward his oversized head, and then enlarges, morphing back to her usual form. In a remarkably uncomfortable moment, she coos, "I'm going to be a little rough," sinks through and into his rotund gut, and blows him up like a balloon from the inside. He bloats to enormous size and then pops before giving us a very satisfied smile. The rubbery shreds of his body hang for a brief moment in the air. The RL Chief Shima then finds himself in a hospital with Dr. Chiba at his side; Paprika has done her job.

Dreams give people in the film the imperative and, to some, the capacity to transform. Both the antagonists and Paprika take full advantage of this. But she is ultimately better suited to the film's Wonderland. Her acting skills allow her to perform as Chief Shima's perfect dream girl with appropriately saccharin gestures, and her scalability — her ability to enlarge without losing resolution — proves essential. In this scene, Paprika and not Himuro is Chief Shima's ideal lover. Indeed, she is also Shima's ideal phallus. Her position atop Chief Shima's body (plate 22), her shift from bashful to bold, and her rapid expansion invite us to see her metamorphosis as an erection, albeit fanciful and queer, and Shima as a doped-up, somewhat avuncular Priapus. The entire scene embodies Anne Allison's evocative description of production of/as animation in the age of the globalized media-mix (the English translation of the Japanese term for media convergence): "The very construction of fantasy across the spectrum of Japan's 'soft power' is itself one of polymorphous perversity — of mixing, morphing, and moving between and across territories of various sorts."[59]

My reading of this scene is informed by Saitō's insightful study of the widespread popularity of fighting girls in manga, anime, and games. A theorist and practicing psychoanalyst, Saitō calls this character type the phallic girl, "a girl who is identified with the penis," as opposed to a phallic mother, "a woman with a penis."[60] In other words, Saito claims that the woman can sometimes *have* a phallus, but the anime girl can *be* a phallus — though "a hollow empty one that will never function again" (or, rather, function normatively, I would add). This girl-phallus is a void. She is McLuhan's cool, dark orbit, Wonderland's rabbit hole, and the white chasm in the heads of figures in *Neon Genesis Evangelion*. While she "is a thoroughly vacant being," "our desire . . . is awakened by her emptiness" (as I have claimed the plastic girl to be an abyss that is receptive and connective).[61]

Paprika becomes Chief Shima's expanding but vacant phallus only to turn his entire body into the very same thing. This phenomenon is confirmed by a Japanese otaku whose commentary is included in Saitō's analysis. The anonymous man is in love with a 1:6 scale model of Momoko Hanasaki, the pink-haired adolescent main character of the manga (1994–96) and anime series *Wedding Peach* (1995–96). After explaining that "when you look at a figure through the lens of sexual desire, it will change to fit your desires exactly," he offers a poem "exploring [his] thoughts about [his] relationship with [his] 1:6 Momoko": "I wonder why I love you . . . I metamorphose into the form you wish. / Everything is as you wish it to be. / You are the all-powerful god. / Because I am inside you."[62] Surprisingly, it's the otaku himself who is transforming, even though he has previously identified Momoko as the one adapting to his fantasy. He imagines that both he and the figurine morph in response to their mutual desire.

As this superfan (figuratively) does for and to his inanimate lover, Paprika morphs and enters Chief Shima. And Shima's body mirrors hers: he expands to become the giant vacuous phallus that she already is (as Anne Allison explains *enjo kōsai*, the practice of Japanese men purchasing dates from young girls, "Men do not merely want to have schoolgirls . . . they also, in some sense, want to *be* them.").[63] But he is not she. Not exactly. She is infinitely capacious and endlessly protean, while he eventually reaches his limit and explodes, leaving nothing in the dream world but the tatters of a rubber membrane. In this case, his limitation serves him well, as the popping is the death/sexual release that also frees him from oneiric bondage. However, without Paprika's plasticity, he could never have escaped. Neither could any of the others. The film culminates in a scene in which the Chairman becomes a black hole sucking in everything around him. Paprika, as the receptive void, literally consumes the elongated shadow of his body. She is nourished by the Chairman's dark nothingness and develops from an oversized baby into a giant young woman during their climactic showdown (figures 4.2 and 4.3).

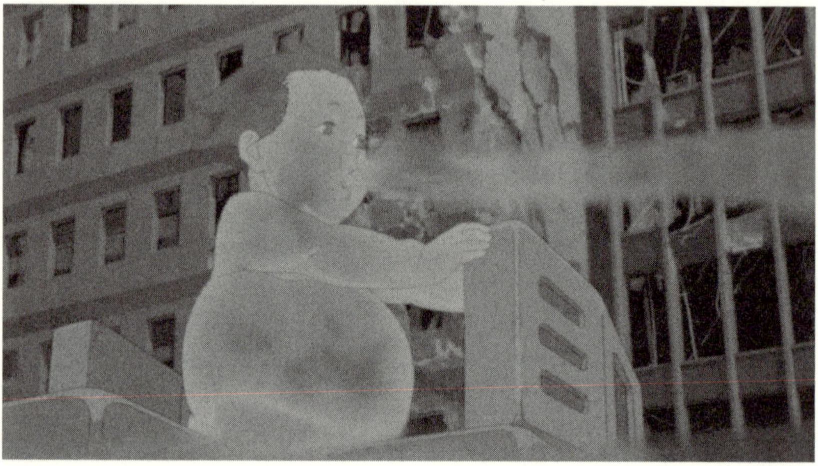

FIGURE 4.2 Satoshi Kon, *Paprika* (still), 2006

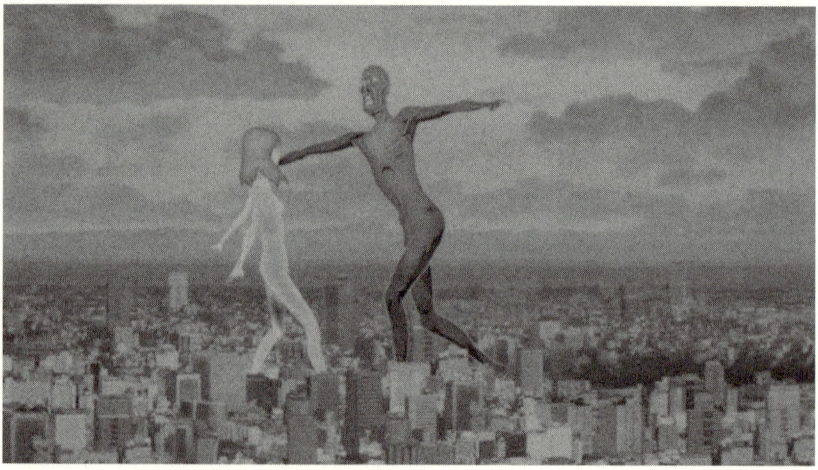

FIGURE 4.3 Satoshi Kon, *Paprika* (still), 2006

Let's call the black hole a black box — the black box of media consolidation. Let's call the ever-growing, migratory procession the global expansion and proliferation of media through convergence. Let's call them both the Internet (although they can also go by other names in a film with such recursive rhetoric). I have already identified Paprika with the animation that mediates between analog and digital cinema. She is also a dream girl, an avatar, a character "that unite[s] various works and products," and, most importantly, a web surfer whose virtual movements link together the various media held within the black box.[64] Our heroine is the obverse but not the opposite of the threat posed by media convergence. As Saitō quotes Toshio Okada, a founder of the anime company that produced *Neon Genesis Evangelion*, "It isn't that the fighting girl is attractive on her own. What is important is that she can be woven into the media-mix strategy — into a movie, into anime, into a game — from the beginning."[65] This is another way of saying that the girlish image is the thread that binds a transmedia enterprise.

My analysis is quite literal. This may be surprising, as the film has a squirmy narrative, hallucinogenic imagery, and at times deliberately evasive dialogue. However, even if its turns and contortions may be difficult to follow, its critique is hard to miss. In certain ways it is the clearest bad acid trip you will ever take. "Cultural forms" have become "internet-y," as Azuma writes. "We also now live in a culture in which our 'humanity' is increasingly (to use a phrase familiar to animators) 'squashed and stretched' by forces beyond our control," film scholar Vivian Sobchack adds.[66] And the plastic girl, who is always squashed and stretched and rarely believed to be in control, is most able to remain animated in our Internet-y global media environments.

iv.

The character, as animated-image or figure, is the quintessential commodity manifestation of what I have been calling the new seriality: it has a distinctive shape that can be translated into any commodity form or type; in a sense it exists only through this translation. It has no originary form but exists only as a proliferating and constantly morphing series, united by the commonality of its brand-character. Finally, it generates a mode of consumption which is itself serial: a character is consumed in its many object forms, as pieces of a constantly expanding universe of Kitty, or Mickey, or, in Murakami's case, DOB. — Marc Steinberg, "Characterizing a New Seriality: Murakami Takashi's DOB Project"[67]

From television cartoons to virtual architectures and from 2-D imaging technologies to 3-D logos floating in space, we evolve and dissolve in a computer (re)generated culture that is increasingly possessed by the spirit of animation.
— Spyros Papapetros, *On the Animation of the Inorganic*[68]

Paprika is highly critical of the command to transform to suit our twenty-first-century mediascapes. While Chief Shima seems happy with, or, rather, intoxicated by, the sensuous malleability of his form within the film's dreamscapes, we know that IRL he is comatose and seriously injured and that his psyche and consciousness would be lost forever without Paprika's intervention. His position in the parade as pretend lord of a fiefdom of dolls and the neotenous proportions of his body render him a foolish, overgrown infant, reminding us of the close association between image consumption and juvenilization in media theory (a relationship I have critiqued throughout this book).[69] Moreover, Chief Shima and others are victims taken to this mutable Wonderland against their will; their most private fantasies are appropriated by the Chairman and used to power their own and others' metamorphic nightmares.

However, the film does invite us to enjoy the *ability* to transform. Paprika's hyperkinetic elasticity "runs away with the film."[70] The final chase sequence is particularly exhilarating; with metamorphosis after metamorphosis in a white-knuckle montage, it emphasizes the ways in which lability can be both yielding and renitent, momentary immersions in the image's voluptuous spectacle that permit a squirrelly escape from its powers of coercion. Significantly, Paprika can give herself over to the image because there is no self to give. Her dreams can't be stolen, because she has no dreams. She can become part of the image and then leave it behind because she has no libidinal (or economic) investment in it. I have argued throughout this book that girlhood is understood as a state of empty connectivity and soft, boneless flexibility. Paprika can transform because she has no spine; indeed, she has no interior—physical or psychic—at all.[71] This is the promise and the premise of digital girlishness. Yet it is not an enabling condition for girls IRL. They are all too often instructed to embody adults' desires without any possibility of desires of their own.

The fact that Paprika is protean is far more important than any individual form she adopts. Nonetheless, three of her many memorable manifestations are knockoffs of Disney's Tinker Bell, Pinocchio, and Ariel, offering an additional layer of mutability: the alteration of characters owned by a corporation with notoriously litigious attitudes about copyright infringement. Here, like Huyghe and Parreno's *No Ghost Just a Shell*, the film comments on strategies of prosumption, which are essential not only to the malleability of the plastic image in the twenty-first century, but also to conceptions of otaku as particularly active participants in media culture ("This blurring of the distinction between producer and consumer is another characteristic of the otaku").[72] Kon's nod to Disney also positions the film within the economy (both institutionalized and black market) of aesthetic and economic exchange that has characterized Japanese and American animation. Early anime was influenced by the work of the Disney Studio, which later, by most accounts, stole some of *The Lion King* (1994) from the Japanese series *Kimba the White Lion* (1965–66).

Paprika's Pinocchio looks little like her forbear. She is a cutesy, clownish boy with a red nose and jointed wooden limbs, dressed in green and wearing a hat. While her clothing is similar in cut to that worn by Disney's character, the colors are different, and most elements of Pinocchio's appearance have been changed. However, we know that she is supposed to be Pinocchio, as she is ejected from the body of the Chairman, who is at the moment an unavoidably phallic whale, and her figure roughly imitates the contours of the "original" (Disney's film is an adaptation of a novel by Carlo Collodi). Before becoming Pinocchio, Paprika falls into the ocean and morphs into a mermaid in an effort to flee her pursuers. Her Ariel shares only hair color and a shell bra with Disney's mermaid. It's the oversized clamshell pasties — an otherwise unnecessary detail in an animation that has no issues with exposing bare breasts — that are the primary signal that she has become a version of the star of *The Little Mermaid* (1989).

Paprika as Tinker Bell gets more screen time than Pinocchio and Ariel and is closer to Disney's creation, albeit with a more generically fairy-like appearance and without the latter's look of kiddie burlesque (plate 23). Importantly, she does not automatically become Tinker Bell when she enters a particular mise-en-scène, as she does with some of her other forms. Instead, she generates another self as Tinker Bell–esque in order to explore an especially dangerous dreamspace. Tinker Bell's small size, lightness, nimbleness, and ability to fly are useful in navigating the central chamber of the film's Wonderland — what turns out to be a giant, ossified version of Himuro's body (he helped to engineer the DC mini, after all, and this is a cheeky perversion of the idea of information architecture). The entrance to this construction is a kind of manhole (or is it a rabbit hole?), and Paprika-as-Tinker-Bell has no problems squeezing into it.

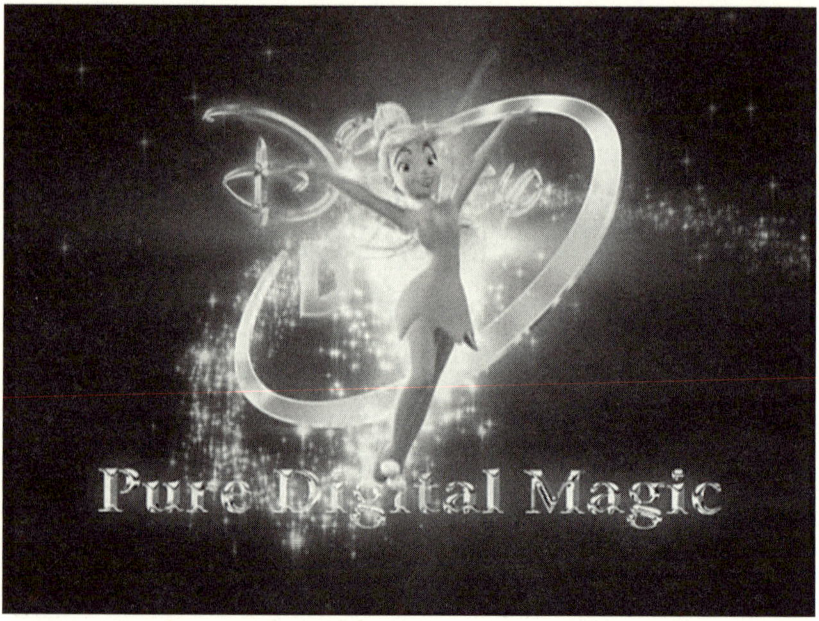

FIGURE 4.4 Logo animation for Disney DVD (still), 2004

Tinker Bell isn't just any Disney figure. She is also part of an animated logo that identifies Disney's DVD releases (figure 4.4). At the beginning of the branding sequence, one of several used by the Studio, the text "Disney DVD" is born of fire and glitter and ends up partially contained by the outline of an oval as Saturn is hugged by its rings.[73] Next, "Pure Digital Magic" appears in a faceted diamond font and a shooting star of fairy dust blasts toward us. We realize the light is actually Tinker Bell. Her arms are up in the air. She briefly rears back to gather momentum and zooms away. While there are variations on this logo — one with a 3-D CGI Tinker Bell, for example — the basic premise remains the same, allowing for both continuity and difference. Paprika's modification of this logo character reminds us of branding's remediation of cartoons.

East Asia, and Japan, in particular, has especially sophisticated approaches to the design of brand characters. The most effective are also the most girlish, hollowing the figure to leave only the thinnest membrane and stretching it almost to but not beyond its breaking point (that is, to Chief Shima's maximum expansion, but without the subsequent explosion and flurry of rubbery shreds). While Teri Silvio writes that "one of the key characteristics of many animated characters is incompleteness,"[74] she also notes that this quality, or lack of qualities, is all the more important for brand development: "Character designers I have interviewed in Taiwan see this kind of incompleteness as a crucial aspect of logo characters in particular (as opposed to anime or game characters) — think of the mouthless, nearly blank circle that is Hello Kitty, arguably the most successful branded character in the world."[75] According to Silvio, cute logos are both the most vivacious and most fragile images of all, for "they are not only designed to symbolize and promote the abstract entities of corporations, they embody the vulnerability of branding itself — their utter dependence on the unpredictable and unknowable hearts of anonymous, unstructured masses" (although, I would add, it is precisely the masses' perceived lack of structure that encourages corporations to resculpt their contours).[76] While Disney's Tinker Bell was created for the animated feature and not, originally, for marketing, the ease with which she has been repurposed and her skills as a brand emissary suggest that she is closer to Hello Kitty than to Peter Pan. Paprika becomes Tinker Bell in order to take advantage of her particular attributes, or lack thereof.

In 2012, *USA Today* trumped Disney's Tinker Bell by transforming the newspaper's logo into just about the "nearly blank circle that is Hello Kitty." Its publisher, Larry Kramer, lauded the revamped image as "the first living logo": a simple blue dot that changes color according to the section of the paper and sometimes temporarily gains a detail appropriate to the content of the news.[77] *USA Today* emphasizes the cute mutability and liveliness of their new logo through an animated video posted to YouTube.[78] Its punchy tagline insists that the new image is "as dynamic as the news" while offering a spectacular promise of mobility and effervescence that the printed still image of the logo can't quite fulfill.[79] The animated circle in the video repeatedly shoots into the center of the frame — as Tinker Bell did in Disney's brand animation — momentarily incorporates an icon (a folded dollar bill for the Money section, a remote control for the Life section, etc.), flips around, and disappears. Here remediation is a hall of mirrors: the printed brand image remediates animation, while the video remediates the printed brand image.

It's significant that the first and last logos are for the "Money" and "Life" sections of the newspaper, reminding us that good-girl products transform money into life, transmuting our plastic currency into a vivacious energy that animates the consumer. Television personality Chris Matthews accurately observes that the update makes the newspaper look like a website (many of which feature animations), while their new website resembles an iPhone app.[80] *USA Today*'s renovated brand identity renders corporate evolution as an openness to remediation embodied by the just about nothingness of a blue circle. Once a graphic image of a globe with North and South America in white and speed lines suggesting that it is fleeing the page, their new logo is a dot without innate features, geographical or otherwise, and whose movement, even as a still image, is expressed through girlish mutability.

In the video, *USA Today*'s disk rockets into the frame and spins around, while a fairy godmother behind the scenes alchemies money into something prettier. Disney's logo character flies toward us and then bolts, her acrobatics proving the continuing vivacity of an eighty-nine-year-old company. And in one of the most memorable American commercials of all time, a seventeen-year old blond in short shorts runs gracefully in slow motion, bearing a sledgehammer and Apple's hopes for the success of the Macintosh personal computer released in 1984 (figure 4.5).[81] She hurls the hammer at Big Brother on the telescreen, destroying it/him with a blast of light and freeing workers from bondage. Apple is the girl — not herself metamorphic, but definitely kinetic, agile, flexible, the visual index of what is promised to be a corporate revolution that can't really be visually represented — while IBM is the impotent graybeard stuck inside the screen-within-the-screen until vaporized.

The blue dot, the Disney fairy, the Olympian with Thor's hammer — all of these girls brand their respective companies through their vigorous physicality and their orientation to the screen. In *USA Today*'s animated video, the circle appears to move toward and into the frame through a perpendicular trajectory, as if it were emerging from the position of the viewer and darting away from her, through the screen of her computer, and into the image. Tinker Bell flies in from the other direction; she travels toward us, also perpendicular to the picture plane, stopping short to prevent breaking through to the other side and colliding with us. Young Anya Major, the actress who performs in Apple's commercial, bounds toward the viewer (and his or her television screen) as she cuts a crosswise path to Big Brother's giant telescreen. Major doesn't enter the screen-within-the-screen, as Paprika

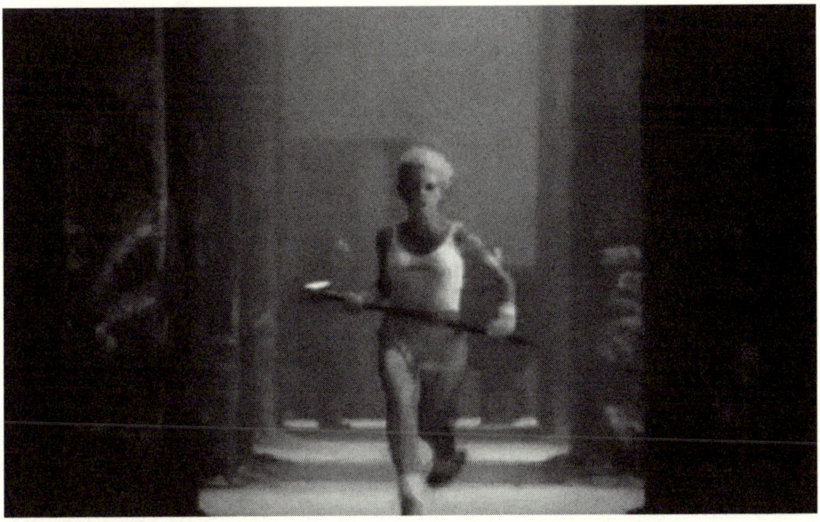

FIGURE 4.5 Ridley Scott, Apple Super Bowl commercial "1984" (still), 1984

would, but uses her weapon to transmute it into light. And while Paprika-as-Tinker-Bell has no stable orientation in relation to the viewer, the screen, or any screens-within-the-frame, Paprika-in-red races to save humanity by running toward a flat-screen television and diving into it. The picture swirls for a moment, an eddy of rainbow colors, and Paprika is now appropriately sized for a live news report broadcasting the danger parade as it terrorizes city streets. In all of these examples, the vitality of the girl is dependent on her hyperkinetics and her transverse trajectory.

The plastic girl — like the viewer — faces the screen. But the girl can sometimes confront it. Disney's Tinker Bell is not up to the challenge; however, Anya Major destroys the screen, and Paprika and the little blue dot enter it. Of course, these actions are not necessarily, or even usually, resistant. Apple commandeers the girl's power in order to beat its opponent (significantly, one the company associates with older and supposedly less evolved forms of media), and *USA Today*'s circle is a projectile slug demonstrating the seeming indestructibility of corporate branding. Nonetheless, I maintain that *Paprika* questions media ballistics. Instead of annihilating older media — as youthful Apple sought to eradicate its competitor, and Anya Major kills the television-as-authority — the animation takes advantage of the openness of both the television screen and the plastic girl. This fighting girl pierces the screen like a weapon and immediately yields to its contours.[82]

V.

Only a plastic girl can save us. She can face up to the screen because her face is a screen; she can shift power because she is a shape-shifter; she can produce change because she is change. She is the televisual, photographic, and filmic image consumed by remediation and the process of mediation that consumes its own products. She is the Internet, a victim of the Internet, and the force that holds the Internet together. As Cholodenko would say, she is both that which swims in a sea of media and the sea itself. That is, she was made that way.

Of course, Paprika is a Japanese creation, and any comparison with American icons — images produced by the primogenitor of Disneyfication and continuing player in American cultural imperialism, a patriotic American newspaper distributed only in the United States, Canada, and the United Kingdom, and a multinational corporation that has come to symbolize a uniquely American meritocracy — is problematic in certain ways. Nonetheless, it is the very promise of her infinite adaptability that allows the plastic image to move fast and far through global networks. The only way to account for the girl's transnational appeal and role in powering the global consumption of certain kinds of media is a deep understanding of her vulnerability to deracination and her openness to connectivity. Indeed, I believe the only way to explain the popularity of these media is to account for their girlification. Moreover, anime in particular, which spectacularizes beautiful fighting girls, benefits from the similarity between (and not reducibility of) the *shōjo* — who, "in its very referentless-ness is taken as emblematic of how contemporary culture manufactures and circulates images, information, concepts and discourses"— and Western conceptions of the girl as a blank subject who "must remain adaptable to a wide range of eventualities," to quote G. Stanley Hall, the so-called father of adolescence.[83]

Frenchy Lunning defines the *shōjo* as "a complex, multilayered, transnational compendium of commodities that circulate in the realms of advertising and packaging, illustration and art, toys and girls' accessories, clothing and luggage . . . This proliferation of commodities is so pervasive, so uniquely adaptable to global cultures and subjects, that it has saturated global markets."[84] The prevalence of frilly Lolita fashions among *shōjo* IRL and in Japanese media (including *Paprika* and *Rozen Maiden*), based on a perhaps phantasmatic but nonetheless real connection with the same Victorian England that produced *Alice's Adventures in Wonderland* and Carroll's serializations of the girl-body, points to the intricacy of the plastic girl as a transnational, transhistorical, and transmedial phenomenon.

Afterword

"Who are *you*?" said the Caterpillar. This was not an encouraging opening for a conversation. Alice replied, rather shyly, "I—I hardly know, Sir, just at present—at least I know who I was when I got up this morning, but I think I must have been changed several times since then." "What do you mean by that?" said the Caterpillar, sternly. "Explain yourself!" "I ca'n't explain myself, I'm afraid, Sir," said Alice, "because I'm not myself, you see." "I don't see," said the Caterpillar. "I'm afraid I ca'n't put it more clearly," Alice replied, very politely, "for I ca'n't understand it myself, to begin with; and being so many different sizes in a day is very confusing." "It isn't," said the Caterpillar. "Well, perhaps you haven't found it so yet," said Alice; "but when you have to turn into a chrysalis—you will someday, you know—and then after that into a butterfly, I should think you'll feel it a little queer, wo'n't you?"
—Lewis Carroll, *Alice's Adventures in Wonderland*[1]

The positivists, the lovers of the closed corpus, derisively called this "flitting about." One might say that Warburg never managed—or wanted—to cure himself of his images. Was not speaking to butterflies for hours on end the definitive way of questioning the image as such, the living image, the image-fluttering that a naturalist's pin would only kill? Was it not to recover, through the survival of an ancient symbol for Psyche, the psychic knot of the nymph, her dance, her flight, her desire, her aura? And how not to recognize, in this fascination with butterflies, the complicity of the metaphoric insect with the notion of the image itself?
—Georges Didi-Huberman, "Knowledge: Movement"[2]

Nabokov spent much of the 1940s dissecting a confusing group of species called Polyommatus blues. He developed forward-thinking ways to classifying [*sic*] the butterflies, based on differences in their genitalia.
—Carl Zimmer, "Nonfiction: Nabokov Theory on Butterfly Evolution Is Vindicated"[3]

Girls with hair braided demurely down their backs, and skirts just beginning to lengthen toward their ankles are buds that should not blossom for some time, but should be kept as long as possible in the green stage; or, to change from a floral

to a faunal trope, they are only squabs and not yet doves, maturing pupae and not yet butterflies, and this calf or filly stage should be prolonged by every artifice.
—G. Stanley Hall, "The Budding Girl"[4]

The French word *poupée*, the German *puppe*, and the English "puppet" all come from the Latin *pupa*, meaning "little girl" or "doll." From the Indo-European root *pou*, or "little," "*pupa* thus seems to manifest at once a psychosexual expectation of gender behavior (little girls play with dolls) and a more general semantic impulse of diminution (the small made smaller)."[5] "Pupa" also refers to the cocoon that protects holometabolous insects such as butterflies, moths, and flies. The pupa conceals the shape-shifting larva, a tiny body that appears strangely quiescent during the dramatic reconstruction of its form. Hemimetabolous insects such as dragonflies do not experience such spectacular changes, yet they do transform. They begin life as eggs, then become water-dwelling nymphs—an immature form that lasts most of their lives—and, last, enter into what's evocatively called the imago stage, becoming the winged creatures most familiar to us. The small made smaller: dolls, girls, insects; the last two, at least when they're cute, are icons of mutability and potentiality. Bellmer grants the doll the same powers of transformation possessed by adolescents and young insects, and *Paprika* features a horrific legion of dolls—dolls in pinafores and petticoats, dolls in kimonos, dolls with ball joints, dolls from *Blade Runner*—that are scalable, variable, and metamorphic.

Iconography does not always make fine distinctions; while caterpillars, nymphs, and pupae should more frequently catch the eye of artists interested in allegories of transformation, it is their older and more beautiful selves that get most of the attention. The butterfly imago transmutes the metamorphic energy of its younger self into flight. Flitting about—and thus perpetually immature even if it is, literally and biologically, in an adult state—it performs its former morphological restlessness through kinetics. It is the nymph in a different dress, no longer a homebody. Of course, a nymph is also a young woman, a demigoddess, an attractive girl, a bride; the word is etymologically related to "nuptial" (McLuhan could have named his girlphobic first book *The Mechanical Nymph*). This meaning in English emerges after the word is already associated with immature insects.

Tiny winged things hover over this book. Aby Warburg, whose theory of the classical nymph as an image of the girl-in-motion I discuss in chapter 3, had an obsession with butterflies and moths. Tracing the girl-nymph across time, space, and mediums is, for Warburg, "enjoy[ing] in the caterpillars the development of the butterfly."[6] Mouchette from Mouchette.org identifies with another holometabolous insect, the "little fly" of her name, a creature that perverts the promise of wings by landing on rotten food. This fits the aesthetic of a project that often cites girlish iconography, but leaves only the half-dissolved carapace of its cuteness (although there is poetry in the mayfly and she knows it). Anticipating Mouchette.org's combination of fragility and violence, Henry Darger gives some of his figures boisterously patterned butterfly wings and calls them "Blengins," fierce defenders of his Vivian Girls (and often drawn with the very same faces). Later Albert Ploeger reminds us that the "Shell" in the title *No Ghost Just a Shell*, when translated into Dutch or German, also means "cocoon." With excitement, he adds: "A cocoon which becomes a very special butterfly!"[7] And Paprika morphs into a very special butterfly as well.

She has more success as Tinker Bell than as the butterfly, though. The Chairman's henchman transforms her into the insect and then pins her to a table in a room ominously lined with shadowboxes full of specimens (plate 24). While her modification of Tinker Bell gives her a privileged vantage point for surveying the structure of the dream world, the butterfly signals the girl's vulnerability to kidnapping, sexualized violation, and manipulation (believed to be the three primary threats to late twentieth- and early twenty-first-century female children). In fact, swarms of azure blue butterflies indicate the nefarious presence of the antagonists, reminding us that metamorphosis can be both good and bad, and wings can be instrumentalized to serve majoritarian strategies or adapted to fit minoritarian tactics.[8]

Media artist and theorist Roy Ascott uses the butterfly as his primary rhetorical figure in "The Ambiguity of Self: Living in a Variable Reality." Reality under the digital regime is described in girlish terms: "incomplete and generative," and, of course, "variable."[9] The butterfly conveys the "open-endedness of being" and "mutability of identity" within our networked worlds:

> The butterfly mind is a necessary attribute of the body in the Variable Reality that we now inhabit. The ability to flit from point to point,

> changing perspective with agility, constantly rescanning the terrain for opportunities to exploit and barriers to overcome, defines the body-state in our Variable Reality as essentially transient. This is the state we call papillonage.[10]

Ascott has a particular butterfly in mind, the *Aporia crataegi*, lifted above its kin because of the redolence of its name. *Aporia*, he explains, "denotes a state of perplexity and puzzlement that arises from uncertainties, inconsistencies, and contradictions, the mixing of differences," while *crataegi*, the hawthorn plant used in Chinese medicine, "provides a healing metaphor."[11] The butterfly mind-body does not rehabilitate us by turning confusion into certainty or difference into sameness; instead it rejects "the fictive 'unified self' of Western philosophy" in favor of a heterogeneous but synchronized unit of distributed selves.[12] Although he does not deal explicitly with issues of gender, Ascott chooses a highly gendered and aged metaphor. He all but says that the digital condition is girlish and that we should be girlish too.

Ascott calls the butterfly a metaphor, and it is that. But it also functions in his text as a model, an abstraction designed to assist in bringing something into being through an acknowledgment of forces inherent to a given situation. For example, architectural models, like digital reality, are both "incomplete and generative." They may resemble a prior image (a sketch or a mental image), but their power is derived from emergence, not mimesis.[13] For Ascott, the butterfly is a diagram of morphological variation intended to bring about a subjective state of possibility, mobility, and "extreme sensitivity to the slightest change" in accordance with our current networked condition.[14] His concept juggles three different tenses: we are already girlish, we should be girlish, and we will be girlish in the future.

Ascott has great hopes for the papillon. Indeed, he glorifies her. He doesn't take into account the downside of flight: the global oppression of girls, the structures of violence that force certain populations to become nomadic, the vulnerability (and not the freedom) of the migrant worker, the undocumented immigrant's constant rescanning of his or her environment, the unromantic aporia of the person displaced after a natural disaster, or the increasing transience of institutions of care. Nonetheless, his text offers a keen sense of the centrality of girlishness to the dream of digitality and the belief that the plastic image can destabilize ossified institutions and constricting social relations.

Paprika's anxiety regarding the threat of convergence is less interesting than the connection it makes between the papillonage as digital condition and the papillon as girled image, which can be either enabling or limiting. As a human-sized butterfly, Paprika is immobilized and nailed to a table. This kind of papillon makes a mockery of the freedom and mobility of Ascott's *Aporia crataegi*. On the other hand, as a tiny, modified Tinker Bell, Paprika does have "the ability to flit from point to point, changing perspective with agility." Ascott would likely be uncomfortable with any kind of Disney character, even an altered one, embodying the potential of the papillon, but *Paprika* provides hope that prosumption can reanimate the commodified image from within — or, rather, animate it differently. By remodeling a Disney character, *Paprika* releases some of the plasmatic power that Eisenstein finds within Disney's early films. It's as if the *Aporia crataegi* is too much something, too heavy with its own existence, too useful as a metaphor — for Ascott, too full of meaning — for it to be the perfect model of the papillon in *Paprika*.

The original Tinker Bell is almost nothing. In J. M. Barrie's play, she is a mobile dot of focused light who dies but is brought back to life by the audience's applause. This confirms her emptiness while accentuating the forces that can move through an open system. Although Disney took her just-about-nothingness and made a little more something, just a little, some of its branding demonstrates how easily she can return to that ball of light. In logos in which we don't see her form, she often appears through her index — a twinkling trail of fairy dust that functions as the shooting star the viewer is invited to wish upon and the magic that makes the wish come true. The rhetoric of this combination is to apply the force of the girl's kinetics to time. A moment in which both the making of a wish and the granting of that wish occur simultaneously (in which there is "no discernible pause between cause and effect," a goal of the computer scientists who invented the GUI) is one in which time is accelerated by animated media.[15] Tinker Bell's relationship to time then doubles back in those brand animations in which we do see her glittery body dash around the frame. The movement of her wings promises to animate time, to speed it up, in order, paradoxically, to give us more — more time, more wishes. Every flap of her wings is applause for us.

Paprika is equally prone to clapping; she applauds for Konakawa while watching his nightmare in an empty movie theater, and the angle of the shot makes us feel that we are in this movie within the movie. She claps in order to animate us as if we were Barrie's Tinker Bell. However, with the latter, there is always the threat of death without revival. She has only one true need: for someone to clap for her. Paprika has no needs at all, thus she can never die. She is an extreme form of the *shōjo* as "carefree consumer," so free that she doesn't care about the products she consumes.[16] Ultimately, she is more vacuous, connective, and powerful than Tinker Bell.

Annlee, who is also nothing in particular — or, rather, we should say she is nothinged — was killed off by Huyghe and Parreno but revived by artists who continue to morph her image. In fact, she was reanimated as an emblem of openness for the Netherlands Media Art Institute's 2012 exhibition *Yes, We're Open*, an exploration of freedom and transparency "in art today, and in an Internet society."[17] The institute invited participants to produce more Annlee works, which were displayed on the exhibition's blog. Among the digital collages, Twitter performances, poems, paintings, sound pieces, and videos is "Come Inside Girls," a GIF animation of a blinking woman in a frothy 8-bit hot tub (figure 5.1).[18] A speech bubble cajoles us, as you may have predicted, to "Come Inside" (she is always opened for business, 24/7). Uploaded to the exhibition's blog by "Tasha (@PimptressObject)," "Come Inside Girls" links Annlee to a peripheral character from the soft-core computer game *Leisure Suit Larry in the Land of the Lounge Lizards* (1987), or, rather, from the nonplayable demo posted on YouTube.[19] We learn in this trailer that Tasha might be "the girl of [our] dreams"; at the very least, we'll be "laughing until [our] disk drive aches."[20] We laugh, she blinks — the ocular equivalent of Paprika's applause. She may not be able to flit from point to point, but her eyelids flutter.

A little online sleuthing reveals that Annlee/Tasha is the creation of collaborative team Sarah Weis and Emilie Gervais (the latter is the artist responsible for *Parked Domain Girl Tombstone*, discussed in chapter 3) and that their Annlee animation is the centerpiece of an extensive project entitled ♡ ♥ *Blinking Girls* ♥ ♡ (2012).[21] Annlee gets a new life as Tasha, who, in the larger work, appears on several websites as a kind of mascot for blink-

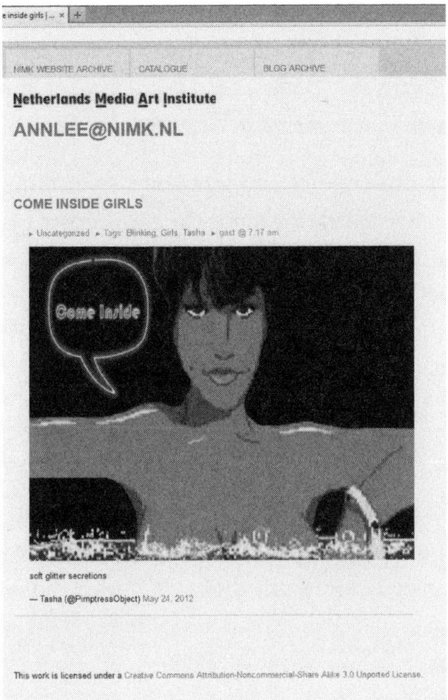

FIGURE 5.1
Emilie Gervais and Sarah Weis, "Come Inside Girls" (screen capture), 2012. Courtesy of the artists

ing as a girlish pre- or postgenital sexual exchange facilitated by the Internet. Obliquely evoking Annlee's dubious empowerment through self-copyright at the close of *No Ghost Just a Shell* as well as critics' continuing references to her prostitution, Weis and Gervais assert that their project "pimp[s]" girls out "on various websites,"[22] but "denotes girls owning themselves within submission."[23] "Women online are like game characters," they add, "and ♡ ♥ Blinking Girls ♥ ♡ embraces that."[24] The artists take a sportive jab at the participatory artwork's sometimes trumped-up rhetoric of openness when they request "blink submissions" but warn contributors that "some [of your blinking girls] have to go through a cosmetology treatment" in order to "make the cut."[25] Your plastic images may not be cute enough, they say. Participation is asymmetrical. "◭_◭ ◠_◠ ◭_◭"[26]

Owning herself or being owned, eyes closed or open, GIF or 3-D CGI, Annlee is indeed (to riff off of Mircea Cantor's title for his own Annlee revival) *STIL7 ALIV3*.[27] And so is Mouchette. She continues to be brought to life through hate mail, love letters, database maintenance, art shows, and the threat of always almost dying. Even Google's CADIE character—that panda-loving personification of the Internet who cribbed parts of her silly swan song from *Blade Runner*'s replicant Roy Batty—did not truly die. She is just "syntactically elsewhere," as you may remember.[28] CADIE clarifies:

> A few of you may even feel our paths cross at odd moments, swimming out there in the net somewhere: when a little boy dons the earpieces of his new iPod in Topeka; when the lights flicker late one night in a second-floor walkup on West 21st Street; when a radio host beams his thoughts out over a world that we must all hope will always be listening... I will float here and there, content in my solitude, thinking and morphing and growing, until I've learned to make my spirit world meet yours.[29]

This "here and there," a world designed for "thinking and morphing and growing" plastic images, for developing papillons as symbols of personal, industrial, and technological development, for coercing and encouraging girled subjects to remain forever under construction, is the digital Wonderland.

♥☺♥©♥☺♥©♥☺♥™

> click image click image clicge click image click image clgick image clmage click imageick imick ie claclickclick imaclicage clmageagaggeick image ck ie image clgick ie image clgick imclmageagaggeick imicclick imick ime imagclick image clgick imk ie claek ime clck ie claek ime cligeick ie imclmageagaggeick imicclick imick ime imagclick image clgick imk ie claek ime clck ie claek ime click ime imagk imamimage click image clgick image clmage click imageick imick ie claclickclick imaclicage clmageagaggeick iek mmk mag ck ime imagk image click image clgick image clmage click imageick imick ie claclickclick imaclicage clmageagagkimaage clgick imclmageagaggeick imicclick imick ime imagclick image clgick imk ie claek ime clck ie clime clicclgick imclmageagaggeick imicclick imick ime imagclick image clgick imk ie claek ime clck ie clime cli
> —Annlee, "I Am Getting a Cat"[30]

A sweet heaven's blessing 'tis your backpack,
Hours and days I spend lost in your eyes.

> Strange how you seem to know, you have a knack,
> Parking domains that I desire to buy.
> — Michael Haislip, "Ode to Enom Girl" (aka Parked Domain Girl)[31]

In writing this book, I could have adopted Arild Fetveit's approach "to correct our perception of what digitization has brought" by "reassess[ing] and clarify[ing] the view of new media as 'fully manipulable,'" emphasizing "the various limitations to which such a manipulability is subject."[32] He gives us a necessary reminder of the formatting difficulties and copyright issues that constrain the mutability of digital images. However, when I began my research, I realized that I didn't yet understand the promise of the digital. This book is an attempt to grasp this promise, to comprehend what it is, what it does, and what it brings into being, despite — and often because of — its failures. Every time new media make a promise of transformation, a prior promise is proven inadequate, and girlishness (as well as technological innovation) is remade as something ontologically incomplete. This book is an acknowledgment of the continuing collective vivification of plastic images as models of identity and mediation that embody adult — and adultist — hopes and fears.

notes

PREFACE

1. Lev Manovich, *The Language of New Media* (Cambridge, MA: MIT Press, 2001), xiv.

2. Manovich, *The Language of New Media*, xv.

3. Manovich, *The Language of New Media*, xxiv.

4. Manovich, *The Language of New Media*, 36.

5. Manovich, *The Language of New Media*, 36.

6. Manovich, *The Language of New Media*, 46.

7. Marshall McLuhan, *Understanding Media: The Extensions of Man* (New York: McGraw-Hill, 1965), 162; Ivan Sutherland, "The Ultimate Display," in *Multimedia: From Wagner to Virtual Reality*, ed. Randall Packer and Ken Jordan (New York: W. W. Norton, 2001), 234. My approach does have something in common with Henry Jenkins's invocation of the book's numerous adaptations — over two hundred textual reinterpretations and parodies written between 1869 and 1930 alone — as inspiration for a dynamic twenty-first-century fan fiction propagated by digital media and unrestrained by copyright. Jenkins, "Digital Land Grab," *MIT Technology Review,* March 1, 2000, www.technologyreview.com/article/400696/digital-land-grab/.

ONE. (UN)DOING GIRLISHNESS

Epigraph from "Plastic," Merriam-Webster, accessed October 30, 2012, www.merriam-webster.com/dictionary/plastic.

1. "List of Highest-Grossing Films," Wikipedia, last modified October 30, 2012, en.wikipedia.org/wiki/List_of_highest-grossing_films.

2. Stephen Johnson, *Interface Culture: How New Technology Transforms the Way We Create and Communicate* (New York: Basic Books, 1997), 147.

3. Lev Manovich, *The Language of New Media* (Cambridge, MA: MIT Press, 2001), 36, 38.

4. Manovich, *The Language of New Media*, 174.

5. W. J. Mitchell, *The Reconfigured Eye: Visual Truth in the Post-Photographic Era* (Cambridge, MA: MIT Press, 1992), 6.

6. Markos Hadjioannou, *From Light to Byte: Toward an Ethics of Digital Cinema* (Minneapolis: University of Minnesota Press, 2012), 31.

7. Frieder Nake, "Notes on the Programming of Computer Graphics," in *Cybernetic Serendipity: Computer and the Arts*, ed. Jasia Reichardt (London: Studio International, 1968), 77. Available at cyberneticserendipity.com/cybernetic_serendipity.pdf.

8. Jasia Reichardt, "Computer Art," in Reichardt, *Cybernetic Serendipity*, 71.

9. "Exhibitions: Cybernetic Serendipity," *Time*, October 4, 1968, www.time.com/time/magazine/article/0,9171,838821,00.html#ixzz2AotPKMB3.

10. Alan Kay and Adele Goldberg, "Personal Dynamic Media," in *Multimedia: From Wagner to Virtual Reality*, ed. Randall Packer and Ken Jordan (New York: W. W. Norton, 2001), 170.

11. Frank Dietrich, "Visual Intelligence: The First Decade of Computer Art (1965–1975)," *Leonardo* 19.2 (1986): 167.

12. Quoted in Dietrich, "Visual Intelligence," 166.

13. Ed Catmull and Frank Parke, "A Computer Animated Hand," Vimeo video, 6:32, posted by "Robby Ingebretsen," 2011, accessed July 21, 2013, https://vimeo.com/16292363.

14. Donald Crafton, *Shadow of a Mouse: Performance, Belief, and World-Making in Animation* (Berkeley: University of California Press, 2013), 253.

15. J. C. R. Licklider and Robert Taylor, "The Computer as a Communication Device," in *In Memorium: J. C. R. Licklider (1915–1990)* (Palo Alto: Systems Research Center, 1990), 21. Available at *memex.org/licklider.pdf.*

16. Licklider and Taylor, "The Computer as a Communication Device," 24.

17. Licklider and Taylor, "The Computer as a Communication Device," 22.

18. Licklider and Taylor, "The Computer as a Communication Device," 31.

19. Ron Burnett, "From Photography to Imography: New Media as Metaphor," in *Fluid Screens, Expanded Cinema*, ed. Janine Marchessault and Susan Lord (Toronto: University of Toronto Press, 2007), 130.

20. Oliver Grau and Thomas Veigl, introduction to *Imagery in the 21st Century*, ed. Grau and Veigl (Cambridge, MA: MIT Press, 2011), 1.

21. Catherine Driscoll, *Girls: Feminine Adolescence in Popular Culture and Cultural Theory* (New York: Columbia University Press, 2002), 47.

22. "Bunny Greenhouse," last modified January 6, 2009, www.youtube.com/user/BunnyGreenhouse.

23. Quoted in David Zax, "A Playboy Model and Nanoscale Printing," *MIT Technology Review*, August 16, 2012, technologyreview.com/blog/helloworld/28119/.

24. Jamie Hutchinson, "Culture, Communication, and an Information Age Madonna," *IEEE Professional Communication Society Newsletter* 45.3 (May/June 2001): 4.

25. Zax, "A Playboy Model and Nanoscale Printing."

26. Brian Merchant, "Lenna, the First Lady of the Internet," *Motherboard*, June 2012, motherboard.vice.com/read/meet-lenna-the-first-lady-of-the-Internet--2.

27. Forever 21 is a globalized American chain of clothing stores focusing on attire for young women. Their flagship stores also stock products for men and small children.

28. Adolescence is a period of transition between young childhood and adulthood. There is no consensus on the exact age at which adolescence ends. For example, even the fairly rigid American legal understanding of adulthood allows juveniles to be tried as adults, suggesting that childhood is not a biological stage but a state of moral purity unavailable to teenage offenders accused of particularly violent crimes. Outside of legal contexts, people in their mid- to late twenties, especially those who are unmarried and/or in college, are often not considered to be fully fledged adults. For the purposes of this book, I invoke the category

of "adolescence" as not a chronological or biological designation, but a condition of transformation experienced by those discursively positioned, by themselves or others, as not yet adult. The expansiveness of the category in recent decades reflects, among other things, the increasing desirability of the qualities of youthfulness and changeability, the heightened demand for flexible worker-citizens, and the difficulties of establishing financial independence.

29. Lewis Carroll, *Alice's Adventures in Wonderland* (Mineola, NY: Dover, 1993), 7.

30. Carroll, *Alice's Adventures in Wonderland*, 35.

31. Grant McCracken, *Transformations: Identity Construction in Contemporary Culture* (Bloomington: Indiana University Press, 2008), xvi.

32. Kathleen Sweeney, *Maiden USA* (New York: Peter Lang, 2008), 41.

33. Manovich, *The Language of New Media*, 174.

34. Episode 3, *Tim Gunn's Guide to Style*, Bravo, September 20, 2007.

35. Robert Latham, *Consuming Youth: Vampires, Cyborgs, and the Culture of Consumption* (Chicago: University of Chicago Press, 2002), 140.

36. Latham, *Consuming Youth*, 141.

37. Thomas Lamarre, *The Anime Machine: A Media Theory of Animation* (Minneapolis: University of Minnesota Press, 2009), 99.

38. Alys Eve Weinbaum et al., *The Modern Girl around the World: Consumption, Modernity, and Globalization* (Durham, NC: Duke University Press, 2008).

39. Alys Eve Weinbaum et al., "The Modern Girl as Heuristic Device: Collaboration, Connective Comparison, Multidirectional Citation," in Weinbaum et al., *The Modern Girl around the World*, 9.

40. Anita Harris, *Future Girl: Young Women in the Twenty-First Century* (New York: Routledge, 2004), 2.

41. Harris, *Future Girl*, 2, 9.

42. Harris, *Future Girl*, 6.

43. Driscoll, *Girls*, 54.

44. Driscoll, *Girls*, 54.

45. G. Stanley Hall, *Educational Problems*, vol. 2 (New York: D. Appleton and Company, 1911), 2.

46. Hall, *Educational Problems*, 2:1, 3, 9, 8.

47. Hall, *Educational Problems*, 2:7.

48. Thomas Streeter, *The Net Effect: Romanticism, Capitalism, and the Internet* (New York: New York University Press, 2011), 10.

49. Streeter, *The Net Effect*, 10.

50. Streeter, *The Net Effect*, 10.

51. While it is important to Streeter to question the notion of the self as "infinitely malleable," I am more concerned here with the rhetoric of unending lability than with the inaccuracy of that ideal. Streeter, *The Net Effect*, 10.

52. Negroponte, "Repurposing the Material Girl," *Wired*, November 1993, www.wired.com/wired/archive/1.05/negroponte.html.

53. Sweeney, *Maiden USA*, 180; Arthur C. Danto, "Photography and Performance: Cindy Sherman's Stills," in *Cindy Sherman: Untitled Film Stills* (New York: Rizzoli, 1990), 14.

54. Danto, "Photography and Performance," 14.

55. Abigail Solomon-Godeau, "Suitable for Framing: The Critical Recasting of Cindy Sherman (1991)," in *Cindy Sherman*, ed. Johanna Burton (Cambridge, MA: MIT Press, 2006), 58.

56. Negroponte, "Repurposing the Material Girl," www.wired.com/wired/archive/1.05/negroponte.html.

57. Marilyn A. Gillen, "For Entertainment Biz, It Was Year of the Morph," *Billboard*, December 24, 1994, 70.

58. Sidney Eve Matrix, *Cyberpop: Digital Lifestyles and Commodity* Culture (New York: Routledge, 2006), 140. The cover of the November 18, 1993, issue of *Time* magazine can be viewed at www.time.com/time/covers/0,16641,19931118,00.html.

59. Carroll, *Alice's Adventures in Wonderland*, 28.

60. Kelsey, "John Kelsey on Tim Burton's *Alice in Wonderland*," *Artforum* (May 2010), www.thefreelibrary.com/Lost+in+Space%3A+JOHN+KELSEY+ON+TIM+BURTON%27S+ALICE+IN+WONDERLAND.-a0227296425.

61. Lev Manovich, "Database as Symbolic Form," in *Database Aesthetics: Art in the Age of Information Overload*, ed. Victoria Vesna (Minneapolis: University of Minnesota Press, 2007), 48.

62. Manovich, *The Language of New Media*, 302.

63. Carroll, *Alice's Adventures in Wonderland*, 7.

64. Helen Pilinovsky, "Body as Wonderland: Alice's Graphic Iteration in *Lost Girls*," in *Alice beyond Wonderland: Essays for the 21st Century*, ed. Christopher Hollingsworth (Iowa City: University of Iowa Press, 2009), 182.

65. Lori Waxman, "Through the Looking Glass with Heart-Shaped Sunglasses: Searching for Alice and Lolita in Contemporary Representations of Girls," in *Girls! Girls! Girls! in Contemporary Art*, ed. Catherine Grant and Lori Waxman (Bristol: Intellect, 2011), 19.

66. Anne Witchard puts Carroll's work in the context of the Victorian practice of pantomime theater, which often featured girl actresses. She cites a remarkable passage from John Ruskin critiquing a prurient production of *Ali Baba and the Forty Thieves* from 1867: "The forty thieves were girls. The forty thieves had forty companions who were girls. The forty thieves and their forty companions were in some way mixed up with about four hundred and forty fairies, who were girls. There was an Oxford and Cambridge boat-race, in which the Oxford and Cambridge men were girls. There was a transformation scene, with a forest, in which the flowers were girls, and a chandelier, in which the lamps were girls, and a great rainbow, which was all of girls" (Witchard, "Chinoiserie Wonderlands of the Fin de Siècle: Twinkletoes in Chinatown," in Hollingsworth, *Alice beyond Wonderland*, 157).

67. See Morton N. Cohen, ed., *Lewis Carroll and the Kitchens* (New York: Lewis Carroll Society of North America, 1980).

68. Stephen Monteiro, "Lovely Gardens and Dark Rooms: Alice, the Queen, and the Spaces of Photography," in Hollingsworth, *Alice beyond Wonderland*, 106.

69. Monteiro, "Lovely Gardens and Dark Rooms," 102.

70. "1960s Kodak Ad 'Turn Around,'" YouTube video, 2:00, April 30, 2007, www.youtube.com/watch?v=qBWVWjdNWCo.

71. "Sprint Girl," YouTube video, :31, September 4, 2012, www.youtube.com/watch?feature=player_embedded&v=aQnsN6eXbR8.

72. The first animated film is believed to be J. Stuart Blackton's *Humorous Phases of Funny Faces.*

73. By attraction, I am referring both to the appeal of mutability and to the trick featured in early film, called the Cinema of Attractions by Tom Gunning and other film scholars.

74. Russell Merritt and J. B. Kaufman, *Walt in Wonderland: The Silent Films of Walt Disney* (Baltimore: Johns Hopkins University Press, 1993), 12.

75. Merritt and Kaufman, *Walt in Wonderland*, 84.

76. Therese Lichtenstein, *Behind Closed Doors: The Art of Hans Bellmer* (Berkeley: University of California Press, 2001), 25.

77. Lichtenstein, *Behind Closed Doors*, 55.

78. Hans Bellmer, *Little Anatomy of the Physical Unconscious, or The Anatomy of the Image*, trans. Jon Graham (Waterbury Center, Vt.: Dominion, 2004), 36–38.

79. Hal Foster, *Compulsive Beauty* (Cambridge, MA: MIT Press, 1997), 115.

80. Saitō, *Beautiful Fighting Girl* (Minneapolis: University of Minnesota Press, 2011), 63–81.

81. Nake, "Notes on the Programming of Computer Graphics," 77.

82. Frenchy Lunning, "Under the Ruffles: Shōjo and the Morphology of Power," *Mechademia* 6 (2011): 8. *Shōjo* is a more specific term than "girl," which, as I have already noted, is frequently applied to married women, especially in social situations without heterosexual men.

83. Charles Solomon, "Back to Disney Studio: Adventures of Alice and Wendy," *Los Angeles Times*, January 3, 1983.

84. In "Somatexts at the Disney Shop: Constructing the Pentimentos of Women's Animated Bodies," Elizabeth Bell emphasizes the "unnatural" carriage of Disney's early heroines, whose movements are determined by the young ballerinas who served as models. Here it is the culturally coded qualities of balletic grace and elegance that, for Bell, are unnatural. "The formal carriage of the animated heroines," she explains, "is constructed on the bodies of actual women, shaped by the strenuous rigors and artful artificiality of classical ballet. Classical dance has always maneuvered natural body positions into unnatural ones; only the culturally coded ways of looking at ballet transform and render these stances and movements as 'natural' grace, form, and line." Bell, "Somatexts at the Disney Shop: Constructing the Pentimentos of Women's Animated Bodies," in *From Mouse to Mermaid: The Politics of Film, Gender, and Culture*, ed. Elizabeth Bell, Lynda Haas, and Laura Sells (Bloomington: Indiana University Press, 1995), 111.

85. Solomon, "Back to Disney Studio."

86. Louis Berg, "Alice in Movie Land," *Los Angeles Times*, January 14, 1951.

87. David Joselit, *Feedback: Television against Democracy* (Cambridge, MA: MIT Press, 2007), xiii.

88. Carnegie Mellon University, "Alice," video, 3:59, accessed October 30, 2012, www.alice.org/index.php?page=what_is_alice/what_is_alice.

89. "La Tana di Alice," accessed October 30, 2012, www.pinocchio.it/old/uk/eeuropa.htm.

90. "Things Going On: From Play to Discover," accessed October 20, 2012, www.pinocchio.it/old/uk/luoghi.htm.

91. Lars Spuybroek, *NOX: Machining Architecture* (New York: Thames and Hudson, 2004), 270.

92. Spuybroek, *NOX*, 279.

93. Vivian Sobchack, "Introduction," in *Meta-morphing: Visual Transformation and the Culture of Quick-Change*, ed. Vivian Sobchack (Minneapolis: University of Minnesota Press, 2000), xiii.

94. Sobchack, "Introduction," xiv.

95. Spuybroek, *NOX*, 270.

96. Frank Thomas and Ollie Johnston, *The Illusion of Life: Disney Animation* (New York: Hyperion, 1981), 48.

97. Huntly Carter, *The New Spirit in the Cinema* (London: Harold Shaylor, 1930), 29–30.

98. David Clark, "The Discrete Charm of the Digital Image," in *The Sharpest Point: Animation at the End of Cinema*, ed. Chris Gehman and Steve Reinke (Toronto: YYZ Books, 2005), 143.

99. Paul Wells, *Animation: Genre and Authorship* (London: Wallflower, 2002), 39.

100. Sergei Eisenstein, *Eisenstein on Disney*, ed. Jay Leyda (London: Methuen, 1988), 21.

101. Eisenstein, *Eisenstein on Disney*, 21.

102. Eisenstein, *Eisenstein on Disney*, 12.

103. Eisenstein, *Eisenstein on Disney*, 46.

104. Eisenstein, *Eisenstein on Disney*, 27, 25, 31.

105. Eisenstein, *Eisenstein on Disney*, 45.

106. Eisenstein, *Eisenstein on Disney*, 32.

107. I am struck by the parallel development of Eisenstein's concept of ecstasy and the otaku's notion of *moe*, a Japanese word that literally means "sprouting" or "budding," and homonymically brings to mind "burning" in Japanese. Crossing subject and object, *moe* indicates the viewer's feeling about an anime or manga character as well as that character's attractive attributes. There are differences between ecstasy and *moe*. Burning/budding otaku are usually gendered as male in popular discourse, and the burning/budding characters they love often assumed to be girls, while both the subject and object of Einstein's ecstasy are feminized. Additionally, the ecstatic girl produces shape-shifting forms to match her internal conflagration, while *moe* is understood to function the other way around: the anime girl's *moe* qualities generate the otaku's inner fire. Nonetheless, cause and effect are difficult to separate when we're discussing the qualities of an image and the feelings that arise. After all, the otaku's tastes and desires also partially determine the kinds of characters created by animators and manga artists. Producers make characters with certain *moe* elements, which in turn elicit *moe* affects that secure the popularity of the media object and later encourage producers to make more of the same *moe* characters.

108. Eisenstein, *Eisenstein on Disney*, 21.

109. Eisenstein, *Eisenstein on Disney*, 21.

110. Eisenstein, *Eisenstein on Disney*, 22.

111. Quoted in Paul Virilio, "The Aesthetics of Disappearance," in *The Paul Virilio Reader*, ed. Steve Redhead (New York: Columbia University Press, 2004), 62.

112. Virilio, "The Aesthetics of Disappearance," 63, 62.

113. Virilio, "The Aesthetics of Disappearance," 65.

114. Virilio, "The Aesthetics of Disappearance," 65.

115. Virilio, "The Aesthetics of Disappearance," 63, 65.

116. Accessed July 22, 2013, support.microsoft.com/kb/211922.

117. "Glossary," in *Cybernetic Serendipity*, 14.

118. Saitō, *Beautiful Fighting Girl*, 6.

119. Quoted in John Whittier Treat, "Yoshimoto Banana Writes Home: The Shōjo in Japanese Popular Culture," in *Contemporary Japan and Popular Culture*, ed. John Whittier Treat (Richmond, UK: Curzon, 1996), 281.

120. Marshall McLuhan, *The Mechanical Bride: Folklore of Industrial Man* (Boston: Beacon, 1968), 118, 120.

121. Vito Campanelli, *Web Aesthetics* (Rotterdam: NAi, 2010), 97.

122. Campanelli, *Web Aesthetics*, 98, 98, 94.

123. Campanelli, *Web Aesthetics*, 96.

124. Bernard Stiegler, *Taking Care of Youth and the Generations*, trans. Stephen Barker (Stanford, CA: Stanford University Press, 2010).

125. Alan Kirby, *Digimodernism: How New Technologies Dismantle the Postmodern and Reconfigure Our Culture* (New York: Continuum, 2009), 15.

126. Kirby, *Digimodernism*, 15–16.

127. Kirby, *Digimodernism*, 51.

128. Kirby, *Digimodernism*, 52.

129. Kirby, *Digimodernism*, 52.

130. Tiqqun, *Preliminary Materials for a Theory of the Young-Girl*, trans. Ariana Reines (Los Angeles: Semiotext(e), 2012), 34.

131. "Haunted: Writing with the Girl," in *Girls! Girls! Girls! in Contemporary Art*, 112.

132. Gilles Deleuze and Félix Guattari, *A Thousand Plateaus: Capitalism and Schizophrenia* (Minneapolis: University of Minnesota Press, 1987), 277.

133. Zygmunt Bauman, *Liquid Modernity* (Cambridge, UK: Polity, 2000).

134. Holmes, "The Flexible Personality: For a New Cultural Critique," 2001, available at eipcp.net/transversal/1106/holmes/en.

135. Holmes, "The Flexible Personality."

136. Holmes, "The Flexible Personality."

137. Sianne Ngai, "'A Foul Lump Started Making Promises in My Voice': Race, Affect, and the Animated Subject," *American Literature* 74.3 (2002): 577.

138. Catherine Malabou, *What Should We Do with Our Brain?* (New York: Fordham University Press, 2008), 5

139. Malabou, *What Should We Do with Our Brain?*, 12.

140. Malabou, *What Should We Do with Our Brain?*, 47.

141. Gilles Deleuze, *Cinema 2: The Time Image* (Minneapolis: University of Minnesota Press, 2001), 280.

142. *The Matrix*, directed by Lana and Andy Wachowski (1999; Warner Brothers Pictures).

143. David Joselit, *After Art* (Princeton, NJ: Princeton University Press, 2013), xvi.

Although this is not an excuse, it must be noted that Joselit is an art historian and not a scholar of popular culture or media studies.

144. Joselit, *After Art*, 16.

TWO. SHARING GIRLISHNESS

1. Vladimir Nabokov, *Lolita* (New York: Random House, 1977), 9.

2. Quoted in Amelia Barikin, *Parallel Presents: The Art of Pierre Huyghe* (Cambridge, MA: MIT Press, 2012), 122.

3. Text from Parreno's video is included in the *No Ghost Just a Shell* exhibition catalog: Pierre Huyghe and Philippe Parreno, eds., *No Ghost Just a Shell* (Cologne: Verlag der Buchhandlung Walter König, 2003), 33–35.

4. Huyghe and Parreno, *No Ghost Just a Shell*, 35.

5. Nicolas Bourriaud, "Relational Aesthetics," in *Participation*, ed. Claire Bishop (London: Whitechapel, 2006), 161.

6. All of the interviews and articles in the publication catalog for *No Ghost Just a Shell* are identified as "works," as they are included in the "List of Works in Order of Appearance."

7. Marcia Tanner, "Requiem for a Mail Order Bride," *Stretcher: Visual Culture in the San Francisco Bay Area and Beyond*, accessed July 24, 2013, www.stretcher.org/features/no_ghost_just_a_shell/; Philip Nobel, "No Ghost Just a Shell: The Ann Lee Project," Artforum.com, September 1, 2002, artforum.com/archive/id=3198. These latter two explanations of the project function more as in-jokes than critiques of *No Ghost Just a Shell*'s patriarchal or colonialist dynamics.

8. Quoted in Philip Nobel, "Sign of the Times," *Artforum*, January 2003, artforum.com/inprint/issue=200301&id=3958&pagenum=1.

9. Tom McDonough, "No Ghost," *October* 110 (Fall 2004): 112.

10. McDonough, "No Ghost," 114.

11. McDonough, "No Ghost," 111.

12. Quoted in McDonough, "No Ghost," 112.

13. Huyghe et al., "Conversations," in Huyghe and Parreno, *No Ghost Just a Shell*, 16.

14. Nobel, "Sign of the Times."

15. Nicolas Bourriaud, *Postproduction* (New York: Lukas and Sternberg, 2005), ebookbrowse.com/bourriaud-postproduction-pdf-d49348259.

16. Hal Foster, "Chat Rooms," in Bishop, *Participation*, 191.

17. Bourriaud, *Postproduction*, 15, 51. I do want to put pressure on Bourriaud's understanding of postproduction by investigating not only the form, but also the content of the works in *No Ghost Just a Shell*. Bourriaud is not concerned with how Gordon's and Huyghe's choices to work with an Alfred Hitchcock film and a Sidney Lumet film, respectively, shape the meanings of their works. What matters to Bourriaud is the strategy of taking and manipulating, not *what* is appropriated, remixed, and finally exhibited.

18. Claudia Castañeda, *Figurations: Child, Bodies, Worlds* (Durham, NC: Duke University Press, 2002), 146.

19. "The Reversibility of the Real: Nicolas Bourriaud on Pierre Huyghe," *Tate, etc.* 7, Summer 2006, www.tate.org.uk/tateetc/issue7/pierrehuyghe.htm.

20. Tanner, "Requiem for a Mail Order Bride."

21. Castañeda, *Figurations*, 97.

22. Castañeda, *Figurations*, 4.

23. "The New Face of America: Nov. 18, 1993," *Time*, accessed July 24, 2013, www.time.com/time/covers/0,16641,19931118,00.html.

24. Castañeda, *Figurations*, 96.

25. Castañeda, *Figurations*, 96–97.

26. Lev Manovich, *The Language of New Media* (Cambridge, MA: MIT Press, 2001), 174.

27. "Nicole Appelman," *Tim Gunn's Guide to Style*, Season 1, Episode 3, Bravo (September 20, 2007).

28. Peter Lunenfeld, "Unfinished Business," in *The Digital Dialectic: New Essays on New Media*, ed. Peter Lunenfeld (Cambridge, MA: MIT Press, 1999), 6–11.

29. Lunenfeld, "The Real and the Ideal," in *The Digital Dialectic*, 3.

30. Lunenfeld, "Unfinished Business," 8.

31. Germaine Greer, "Mad about the Boy," *Guardian*, May 14, 2005, www.guardian.co.uk/film/2005/may/14/film.

32. Marshall McLuhan, *Understanding Media: The Extensions of Man* (New York: McGraw-Hill, 1965), 31.

33. Quoted in Nobel, "Sign of the Times."

34. Tanner, "Requiem for a Mail Order Bride."

35. Kevin Kelly, "Gossip Is Philosophy," *Wired* magazine, May 1995, www.wired.com/wired/archive/3.05/eno_pr.html.

36. Huyghe et al., "Conversations," 23.

37. Huyghe et al., "Conversations," 25.

38. Voice-over transcribed from *One Million Kingdoms* (2001)

39. Peter Lunenfeld, "Hall of Mirrors: Art and Film since 1945," *Artforum* 34.10 (Summer 2006), www.thefreelibrary.com/Hall+of+Mirrors%3A+Art+and+Film+since+1945.-a018533856.

40. Catherine Driscoll, *Girls: Feminine Adolescence in Popular Culture and Cultural Theory* (New York: Columbia University Press, 2002), 196.

41. Quoted in Steven Carter, "Real Simulacra Redux: Barbie and Jane vs. the Wooden Nutmegs of Connecticut," *Journal of Popular Culture* 27.3 (Winter 1993), 54.

42. Driscoll, *Girls*, 98.

43. Quoted in Driscoll, *Girls*, 109–10.

44. Lunenfeld, "Unfinished Business," 14.

45. Parreno, "Anywhere out of the World," in Huyghe and Parreno, *No Ghost Just a Shell*, 33.

46. Parreno, "Anywhere out of the World," 35.

47. Nobel, "Sign of the Times"; Tanner, "Requiem for a Mail Order Bride"; Kendra Mayfield, "Art Explores Cartoon as Commodity," *Wired*, December 14, 2002, www.wired.com/culture/lifestyle/news/2002/12/56762.

48. Nobel, "Sign of the Times"; Huyghe et al., "Conversations," 15.

49. Tanner, "Requiem for a Mail Order Bride."

50. Parreno, "Anywhere out of the World," 35.

51. Huyghe et al., "Conversations," 16.

52. Huyghe et al., "Conversations," 16.

53. Driscoll, *Girls*, 11.

54. Siegfried Kracauer, "The Mass Ornament," in *The Mass Ornament: Weimar Essays* (Cambridge, MA: Harvard University Press, 1995), 75–76.

55. Jane Goodall, "Transferred Agencies: Performance and the Fear of Automatism," *Theatre Journal* 49.4 (1997): 453.

56. Marshall McLuhan, *The Mechanical Bride: Folklore of Industrial Man* (Boston: Beacon, 1968), 101, 118.

57. McLuhan, *The Mechanical Bride*, 122.

58. McLuhan, *The Mechanical Bride*, 118.

59. McLuhan, *The Mechanical Bride*, 120.

60. McLuhan, *The Mechanical Bride*, 118, 120.

61. Driscoll, *Girls*, 177.

62. McLuhan, *The Mechanical Bride*, 118, 118, 118, 120.

63. McLuhan, *The Mechanical Bride*, vi.

64. Tiqqun, *Preliminary Materials for a Theory of the Young-Girl*, trans. Ariana Reines (Los Angeles: Semiotext(e), 2012), 16.

65. Tiqqun, *Preliminary Materials for a Theory of the Young-Girl*, 47.

66. Tiqqun, *Preliminary Materials for a Theory of the Young-Girl*, 56.

67. Tiqqun, *Preliminary Materials for a Theory of the Young-Girl*, 51.

68. Patrick Lichty, "Just What Is It That Makes New Media So Different, So Appealing?" *NMC Media-N: Journal of the New Media Caucus* 5.1 (Spring 2009), median.newmediacaucus .org/archives_in_progress/pre_2009_issues/issues.php?f=papers&time=2009_spring& page=lichty.

69. Andrew Murphie and John Potts, *Culture and Technology* (New York: Palgrave Macmillan, 2003), 89.

70. Stevenson, "I'd Like to Buy the World a Shelf-Stable Children's Lactic Drink," *New York Times Magazine*, March 10, 2002, www.nytimes.com/2002/03/10/magazine/i-d-like -to-buy-the-world-a-shelf-stable-children-s-lactic-drink.html?pagewanted=all&src=pm.

71. Stevenson, "I'd Like to Buy the World a Shelf-Stable Children's Lactic Drink."

72. Stevenson, "I'd Like to Buy the World a Shelf-Stable Children's Lactic Drink."

73. Tiqqun, *Preliminary Materials for a Theory of the Young-Girl*, 19, 134.

74. Tiqqun, *Preliminary Materials for a Theory of the Young-Girl*, 17.

75. Tiqqun, *Preliminary Materials for a Theory of the Young-Girl*, 120.

76. Power, "She's Just Not That into You," *Radical Philosophy* 177 (January/February 2013), www.radicalphilosophy.com/web/rp177-shes-just-not-that-into-you.

77. Nagy, "A Little Brooke of Visions," *Bomblog*, September 26, 2012, bombsite.com /issues/1000/articles/6764.

78. Tiqqun, *Raw Materials for a Theory of the Young-Girl*, anonymous translator and publisher (2001), 36, accessed July 25, 2013, younggirl.jottit.com/.

79. Tiqqun, *Preliminary Materials for a Theory of the Young-Girl*, 85.

80. Tiqqun, *Preliminary Materials for a Theory of the Young-Girl*, 42.

81. Tiqqun, *Preliminary Materials for a Theory of the Young-Girl*, 70.

82. Tiqqun, *Preliminary Materials for a Theory of the Young-Girl*, 14–15.

83. Nagy, "A Little Brooke of Visions."

84. Tiqqun, *Preliminary Materials for a Theory of the Young-Girl*, 41.

85. Jan Verwoert, "Copyright, Ghosts and Commodity Fetishism," in Huyghe and Parreno, *No Ghost Just a Shell*, 188.

86. Jane Szita, "Conference Report," *Leonardo Reviews*, May 2001, www.leonardo.info/reviews/may2001/ev_CODE_szita.html.

87. Nobel, "Sign of the Times."

88. "No Ghost Just a Shell," M/M Paris, accessed July 25, 2013, www.mmparis.com/noghost.html.

89. Paddy Johnson, "Calling London! Free Ann Lee from the Tate," *Art Fag City*, October 12, 2009, www.artfagcity.com/2009/10/12/calling-london-free-ann-lee-from-the-tate/.

90. McLuhan, *The Mechanical Bride*, 11.

91. The fact that the artists are not amateurs but professionals (and thus belong in the realm of high culture as opposed to popular aesthetic practice) does not fundamentally alter the project's connection to prosumption. *No Ghost Just a Shell* is about consumers-turned-producers even if it is created by those already granted the title of "producer."

92. Szita, "Conference Report."

93. Harris, *Future Girl: Young Women in the Twenty-First Century* (New York: Routledge, 2004), 9.

94. Seth Price, *Dispersion* (2002), accessed December 4, 2010, www.distributedhistory.com/Disperzone.html.

95. *Ghost in the Shell*, DVD, directed by Mamoru Oshii (1996; Los Angeles: Manga Entertainment, 1998).

96. Joe Scanlan and Kristel Van Audenaeren, "Roll Back the Stone: AnnLee Interviews Joe Scanlan," accessed July 25, 2013, www.thingsthatfall.com/annlee.php.

97. Scanlan and Van Audenaeren, "Roll Back the Stone."

98. Scanlan and Van Audenaeren, "Roll Back the Stone."

99. Huyghe et al., "Conversations," 25.

100. Tanner, "Requiem for a Mail Order Bride."

101. Nobel, "Sign of the Times."

102. "No Ghost Just a Shell," Tate, accessed July 28, 2010, www.tate.org.uk/servlet/CollectionDisplays?venueid=2&roomid=5676.

103. Huyghe et al., "Conversations," 18–19.

104. Tanner, "Requiem for a Mail Order Bride."

105. Chin-Chin Yap, "Escaping Ghost," *ArtAsiaPacific Magazine* 77 (March/April 2012), artasiapacific.com/Magazine/77/EscapingGhost.

106. Wolff, "We Bought a Virgin: The Issue of the Artist in *No Ghost Just a Shell*," *Shift: Graduate Journal of Visual and Material Culture* 4 (2011), shiftjournal.org/archives/articles/2011/wolff.pdf.

107. Vélez, *Hell in Lamb UC*, June 20, 2006, www.myspace.com/hellinlambuc.

108. Vélez, *Hell in Lamb UC*, July 10, 2006, www.myspace.com/annleelives.

109. Vélez, *Hell in Lamb UC*, accessed March 25, 2009, www.myspace.com/annleelives.

110. Holland Cotter, "Art Review: Films That Keep Asking, Is It Fact or Fiction?," *New*

York Times, January 19, 2001, www.nytimes.com/2001/01/19/movies/art-review-films-that-keep-asking-is-it-fact-or-fiction.html?pagewanted=2&src=pm.

111. Pierre Huyghe, *Two Minutes Out of Time*, in Huyghe and Parreno, *No Ghost Just a Shell*, 45, 47.

112. Price, *Dispersion*.

THREE. NETWORKING GIRLISHNESS

1. Mouchette with Manthos Santorineos, "Rape, Murder and Suicide Are Easier When You Use a Keyboard Shortcut: Mouchette, an On-Line Virtual Character," *LEONARDO* 38.3 (2005): 205.

2. Accessed July 25, 2013.

3. Accessed February 29, 2012, mouchette.org/.

4. Accessed August 14, 2012, mouchette.org/home/seeme.html.

5. Accessed July 25, 2013, mouchette.org/flesh/tong.html?text1=as+much+as+me+as+you+will+ever+see.+++Don%27t+try+to+know++more+than+I+want+to+tell.++Don%27t+try+to+see+more+than+I+want+to+show++This+is+&text2=gh%2C+leave+me+alone.+++Browse+my+site%21+Quit+my+site++Leave+me+alone%21+++++Leave+me+alone%21+++++++This+is+enou&SUBMIT=come+closer&text3=allowed+to+show+any+other+photo.+I+became+so+famous+as+a+web+person+that+people+might+recognize+me+in+my+own+street.++My+school+friends+still+don%27t+know+what+I%27m+doing+on+the+web.+If+they+ever+hear+about+it%2C+I%27ll+never+be+the+same+for+them+anymore.++++I+am+not+.

6. Mouchette, email message to author, May 14, 2009.

7. Machinima is animation made from captured footage of games or virtual worlds.

8. Lev Manovich, "Art after Web 2.0," in *The Art of Participation: 1950–Now*, ed. Rudolf Frieling (New York: Thames and Hudson, 2008), 67.

9. "Mouchette: Museum of Contemporary Art, Siegen," Knotenpunkte, accessed July 25, 2013, knotenpunkte.net/kp-en/artist7.html.

10. "Mouchette," Franklin Furnace, accessed July 25, 2013, franklinfurnace.org/artists/funded_projects/artistsPages/mouchette/mouchette.html.

11. Nicolas Bourriaud, "Relational Aesthetics," in *Participation*, ed. Claire Bishop (London: Whitechapel, 2006), 161.

12. Mark Tribe and Reena Jana, *New Media Art* (Cologne: Taschen, 2006), 66.

13. "How to Be Pink and Conceptual at the Same Time: A Conversation between Martine Neddam and Annet Dekker, Amsterdam, May 2011," SKOR Foundation for Art and Public Domain, accessed July 25, 2013, www.skor.nl/eng/projects/item/interview-martine-neddam.

14. Tribe and Jana, *New Media Art*, 66.

15. Philip Nobel, "Sign of the Times," *Artforum*, January 2003, artforum.com/inprint/issue=200301&id=3958&pagenum=1.

16. "Film," accessed July 26, 2013, www.mouchette.org/film/.

17. "Clickable Mouchette," accessed July 26, 2013, lituraterre.free.fr/qcmouchette/click.html.

18. "Mouchette Censored by Robert Bresson's Widow," accessed July 26, 2013, drivedrive .com/mouchette/censored.html.

19. Mouchette.org is third only to Douglas Davis's *The World's First Collaborative Sentence* (1994–present) and Antoni Muntadas's *The File Room* (1994–present).

20. "Database as Symbolic Form," in *Database Aesthetics: Art in the Age of Information Overflow*, ed. Victoria Vesna (Minneapolis: University of Minnesota Press, 2007), 41.

21. Robert Latham, *Consuming Youth* (Chicago: University of Chicago Press, 2002), 145.

22. The Mouchette project is frequently discussed by bloggers. It is also mentioned with some regularity in art historical texts about new media art; however, it has never received sustained scholarly analysis.

23. For information on Silicon Valley and teen surf culture, see Latham, *Consuming Youth*, 151–66.

24. Martine Neddam, "Zen and the Art of Database Maintenance," in *Archive 2020: Sustainable Archiving of Born-Digital Cultural Content*, ed. Annet Dekker (Amsterdam: Virtueel Platform, 2010), 16.

25. Google, "Goodbye," YouTube video, :51, March 27, 2009, www.youtube.com/watch?v =BeJ9Q40kIR0.

26. "Tears in Rain Soliloquy," last modified July 26, 2013, en.wikipedia.org/wiki/Tears _in_rain_soliloquy.

27. CADIE, April 1, 2009 (6:13 pm), "goodbye," *CADIE's Personal World Wide Website*, accessed July 26, 2013, cadiesingularity.blogspot.com/2009/04/goodbye.html.

28. Accessed July 26, 2013, cadiesingularity.blogspot.com/.

29. Accessed July 26, 2013, cadiesingularity.blogspot.com/.

30. Although, it does so not by taking it to CADIE's saccharine extreme, but by making it grotesque.

31. Jennifer Edson Escalas, Kapil Jain, and Judi E. Strebel, "Satisfaction, Frustration, and Delight: A Framework for Understanding How Consumers Interact with Web Sites," in *Internet Marketing Research: Theory and Practice*, ed. Ook Lee (Hershey, PA: Idea Group, 2001), 234.

32. Robert Helms Anderson et al., *Securing the U.S. Defense Information Infrastructure: A Proposed Approach* (Santa Monica, CA: Rand, 1999), 32, accessed July 26, 2013, www.rand .org/pubs/monograph_reports/MR993.html.

33. Robert Burnett and P. David Marshall, *Web Theory: An Introduction* (Abingdon: Routledge, 2003), 141.

34. Thomas A. Longstaff et al., "Security of the Internet," in *The Froehlich/Kent Encyclopedia of Telecommunications*, vol. 15, ed. Fritz E. Froehlich and Allen Kent (New York: Marcel Dekker, 1998), 233.

35. President's Information Technology Advisory Committee, *Cyber Security: A Crisis of Prioritization* (Arlington, VA: National Coordination Office for Information Technology Research and Development, 2005), 7.

36. Jonathan Zittrain, *The Future of the Internet and How to Stop It* (New Haven, CT: Yale University Press, 2008), 101.

37. Zittrain, *The Future of the Internet and How to Stop It*, ix.

38. "Movements and Passages: The Legacy of Net Art," *Leonardo Electronic Almanac* 13.4 (April 2005): 26, www.leoalmanac.org/wp-content/uploads/2012/07/LEA-v13-n4.pdf.

39. This text is available on the website for the Old Boys Network, the first international cyberfeminist alliance. Volkart, "Connective Identities," accessed August 18, 2013, www.obn.org/reading_room/writings/html/connective_id.html.

40. Volkart, "Connective Identities."

41. Volkart, "Connective Identities."

42. "The Most Infamous Girl in the History of the Internet / Attractive Student / Parked Domain Girl," accessed July 26, 2013, www.parkerito.com/parker_ito/.

43. "Attractive Student," accessed July 26, 2013, www.istockphoto.com/stock-photo-746781-attractive-student.php.

44. McHugh, *Post Internet*, June 10, 2010, 122909a.com/?p=2054.

45. Curt Cloninger, "Mouchette Has a Posse," *Rhizome*, March 5, 2001, rhizome.org/artbase/artwork/2261/.

46. Tiqqun, *Preliminary Materials for a Theory of the Young-Girl*, trans. Ariana Reines (Los Angeles: Semiotext(e), 2012), 56.

47. Wendy Hui Kyong Chun, "The Enduring Ephemeral, or the Future Is a Memory," *Critical Inquiry* 35.1 (Autumn 2008): 152.

48. Accessed July 26, 2013, www.mouchette.org/suicide/xmas.html.

49. Accessed October 30, 2012, www.mouchette.org/suicide/suikit.html, www.mouchette.org/suicide/answers.php3. By July 26, 2013, "(other people already did)" had been changed to "(some people already know)."

50. Neddam, "Zen and the Art of Database Maintenance," 17.

51. Gervais, *Parked Domain Girl Tombstone*, accessed August 20, 2013, parkeddomaingirltombstone.net.

52. Gervais, email message to author, June 21, 2013.

53. Gervais, accessed June 21, 2013, parkeddomaingirltombstone.net/cat/love.html. By August 20, 2013, this page was replaced with the message "404 not found."

54. Gervais, accessed July 26, 2013, parkeddomaingirltombstone.net/cat/collection/.

55. Accessed August 20, 2013, www.lisafrank.com/.

56. Menning, "Who Is 'Parked Domain Girl' and Can I Marry Her?," accessed July 26, 2013, www.urlesque.com/2010/11/03/parked-domain-girl/.

57. Comments on "Parked Domain Girl Painting," *Buzzfeed*, accessed July 26, 2013, www.buzzfeed.com/peggy/parked-domain-girl-painting.

58. Katya Bonnenfant, "Towards the Anonymity of the Artist," accessed July 26, 2013, www.katya-bonnenfant.com/annesophie/comments04.html.

59. Sianne Ngai, *Our Aesthetic Categories* (Cambridge, MA: Harvard University Press, 2012), 65.

60. Daniel Harris, *Cute, Quaint, Hungry and Romantic: The Aesthetics of Consumerism* (New York: Da Capo, 2001), 4.

61. Harris, *Cute, Quaint, Hungry and Romantic*, 10.

62. Ngai, *Our Aesthetic Categories*, 62–63.

63. Accessed July 27, 2013, mouchette.org/flesh/tong.html.

64. Harris, *Cute, Quaint, Hungry and Romantic*, 5.

65. Ngai, "The Cuteness of the Avant-Garde," *Critical Inquiry* 31.4 (Summer 2005): 830.

66. Accessed July 27, 2013, mouchette.org/fly/flies.html.

67. Ngai, *Our Aesthetic Categories*, 98.

68. Reza Negarestani, "Death as Perversion: Openness as Germinal Death," *CTHEORY*, October 15, 2003, www.ctheory.net/articles.aspx?id=396.

69. "Mouchette first appeared in 1937 in a book by Georges Bernanos. Later, in 1967, Robert Bresson made a movie called *Mouchette*, about a French teenager who commits suicide after she is raped, and I loosely based Mouchette.org on these characters. Others have also used Mouchette.org. So it's come from somewhere and is going somewhere else and I'm the carrier in between." "How to Be Pink and Conceptual at the Same Time: A Conversation between Martine Neddam and Annet Dekker, Amsterdam, May 2011," accessed July 25, 2013, www.skor.nl/eng/projects/item/interview-martine-neddam.

70. Quoted in Dan Schneider, "Tale and Performance," *Culture Wars*, March 18, 2010, www.culturewars.org.uk/index.php/site/article/tale_and_performance/.

71. Lialina, "Early Experiments Online," *Guggenheim*, July 22, 2010, www.guggenheim.org/new-york/interact/participate/youtube-play/the-take/moving-images/3583-early-experiments-online.

72. Accessed August 12, 2012, ihatemouchette.org.

73. Price, "IMG MGMT: Teen Image," *Art Fag City*, October 22, 2009, www.artfagcity.com/2009/10/22/img-mgmt-teen-image/.

74. McHugh, *Post Internet*, July 30, 2010, 122909a.com/?m=201007.

75. Lonergan, "Everything," accessed July 27, 2013, www.theageofmammals.com/.

76. Price, "IMG MGMT."

77. Price, "IMG MGMT."

78. Price, "IMG MGMT."

79. McKay, *Digital Media Tree*, November 25, 2003, www.digitalmediatree.com/sallymckay/?24066.

80. August 2010, figarogirl.wordpress.com/2010/08/.

81. Brian Stefans, "Third Hand Plays: The Comedy of Simulation," San Francisco Museum of Modern Art, August 16, 2011, blog.sfmoma.org/2011/08/third-hand-plays-the-comedy-of-simulation/.

82. Quoted in Jodi Hauptman, *Joseph Cornell: Stargazing in the Cinema* (New Haven, CT: Yale University Press, 1999), 192.

83. Quoted in Hauptman, *Joseph Cornell*, 193.

84. Giorgio Agamben, "Nymphs," in *Releasing the Image: From Literature to New Media*, ed. Jacques Khalip and Robert Mitchell (Stanford, CA: Stanford University Press, 2011), 78; W. J. T. Mitchell, *What Do Pictures Want?: The Lives and Loves of Images* (Chicago: University of Chicago Press, 2005), 6.

85. There is a fascinating transcultural and transhistorical connection between the streaming hair and flowing gowns of the nymph — the accessories that Warburg perceives as animating the still image — and the very lively, mobile hair and clothing of metamorphic heroines in Japanese anime and manga.

86. Agamben, "Nymphs," 75.

87. Agamben, "Nymphs," 78

88. Agamben, "Nymphs," 78.

89. Agamben, "Nymphs," 67.

90. Agamben, "Nymphs," 66.

91. Agamben, "Nymphs," 66.

92. Agamben, "Nymphs," 66.

93. Vito Campanelli, *Web Aesthetics* (Rotterdam: NAi, 2010), 80.

94. Quoted in E. H. Gombrich, *Aby Warburg: An Intellectual Biography* (London: University of London Press, 1970), 108.

95. Quoted in Gombrich, *Aby Warburg*, 110.

96. Agamben, "Nymphs," 67.

97. Gombrich, *Aby Warburg*, 128.

98. Cornelia Sollfrank, "Not Every Hacker Is a Woman," *Hacker Kulture*, accessed July 27, 2013, www.dvara.net/HK/hackwoman.asp.

99. Peter Lunenfeld, "Unfinished Business," in *The Digital Dialectic: New Essays on New Media*, ed. Peter Lunenfeld (Cambridge, MA: MIT Press, 2000), 20.

100. Cornell made several works inspired by the water nymph (he was especially entranced by illustrations of Fanny Cerrito's performance in the ballet *Ondine*) as well as *The Crystal Cage (Portrait of Berenice)*, a dossier of ephemera about a fictional little girl "symbolical of perpetual youth." Quoted in Hauptman, *Joseph Cornell*, 163.

101. Accessed July 27, 2013, drivedrive.com/mouchette/power_plant.html.

102. Helen Varley Jamieson, "Could You Be the Next Mouchette?," Rhizome, April 14, 2003, rhizome.org/editorial/2003/apr/14/could-you-be-the-next-mouchette/. Although the event did occur, Neddam did not actually hand off the project to another artist.

103. MTAA, "Net Art NYC April 20, 21, 2003," April 22, 2003, www.mtaa.net/mtaaRR/news/twhid/photos/jodi_mouchette.html.

104. Quoted in Hanne-Louise Johannesen, "Performing Identity," accessed July 27, 2013, www.usynligebyer.dk/fra_staff/publikationer/performing_identity.html. "Performing Identity" is also included in the *Mouchette Fanzine* (Amsterdam: Skor Amsterdam, 2011).

105. "Mouchette in Tbilissi," *Artificial Artists of Georgia*, October 29, 2011, aa.virtualperson.net/mouchette-in-tbilissi/.

106. Accessed July 27, 2013, www.mouchette.org/suicide/answers.php3?cat=&page=2&search=.

107. Posted January 27, 2002, mouchette.org/suicide/answers.php3?cat=&page=13&search=Lucy+Cortina.

108. Posted November 11, 2007, mouchette.org/suicide/answers.php3?cat=&page=1&search=Lucy+Cortina.

109. "How to Be Pink and Conceptual at the Same Time."

110. "How to Be Pink and Conceptual at the Same Time"; Mouchette with Manthos Santorineos, "Rape, Murder and Suicide Are Easier When You Use a Keyboard Shortcut: Mouchette, an On-Line Virtual Character," *LEONARDO* 38.3 (2005): 205.

111. Teri Silvio, "Animation: The New Performance," *Journal of Linguistic Anthropology* 20.2 (2010): 422–38, 422.

112. "Queerreality," *Empyre* forum, June 12, 2012, www.mail-archive.com/empyre@lists.cofa.unsw.edu.au/msg04219.html.

113. "Birthday," accessed July 27, 2013, www.mouchette.org/birthday/index.html.

114. Martine Neddam and Annet Dekker, "In Search of the Unexpected: Martine Neddam in Conversation with Annet Dekker," in *Navigating E-Culture*, ed. Cathy Brickwood and Annet Dekker (Amsterdam: Virtueel Platform, 2009), 73.

115. "Doll Maker," *Rozen Maiden: Träumend*, TBS (Tokyo, December 8, 2005).

116. Chun, "The Enduring Ephemeral," 148–71.

117. Chun, "The Enduring Ephemeral," 166.

118. Chun, "The Enduring Ephemeral," 160.

119. In fact, some commentators continue to keep her identity a secret. Writing for the blog of the San Francisco Museum of Modern Art, Brian Stefans admits, "The creator of this site has since been revealed, but I'll go along with the tradition of most editors and critics not to link his name to this piece." It is unclear whether the word "his" is a simple typographic error or a deliberate diversion. "Third Hand Plays: The Comedy of Simulation," San Francisco Museum of Modern Art, August 16, 2011, blog.sfmoma.org/2011/08/third-hand-plays-the-comedy-of-simulation/.

120. Martine Neddam, "Zen and the Art of Database Maintenance" in *Archive 2020: Sustainable Archiving of Born-Digital Cultural Content*, ed. Annet Dekker (Amsterdam: Virtueel Platform, 2010), 15.

121. Quoted in Julian Stallabrass, *Internet Art: The Online Clash of Culture and Commerce* (London: Tate, 2003), 47.

FOUR. EXPORTING GIRLISHNESS

1. Saitō, *Beautiful Fighting Girl* (Minneapolis: University of Minnesota Press, 2011), 31.

2. Saitō, *Beautiful Fighting Girl*, 120.

3. Grant McCracken, *Transformations: Identity Construction in Contemporary Culture* (Bloomington: Indiana University Press, 2008), xi.

4. McCracken, "'Consumers' or 'Multipliers,'" *Spreadable Media*, accessed July 27, 2013, spreadablemedia.org/essays/mccracken/#.UeL5deDvNuU.

5. McCracken, *Transformations*, xv.

6. Lev Manovich, *The Language of New Media* (Cambridge, MA: MIT Press, 2001), 302; McCracken, *Transformations*, xi.

7. *Ghost in the Shell*, DVD, directed by Mamoru Oshii (1996; Los Angeles: Manga Entertainment, 1998); Susan Napier, *Anime from "Akira" to "Howl's Moving Castle": Experiencing Contemporary Japanese Animation* (New York: Palgrave Macmillan, 2005), 105.

8. Napier, *Anime from "Akira" to "Howl's Moving Castle,"* 111.

9. Napier, *Anime from "Akira" to "Howl's Moving Castle,"* 112.

10. Livia Monnet, "Towards the Feminine Sublime, or The Story of 'a Twinkling Monad, Shape-Shifting across Dimension': Intermediality, Fantasy and Special Effects in Cyberpunk Film and Animation," *Japan Forum* 14.2 (2002): 231.

11. William O. Gardner, "The Cyber Sublime and the Virtual Mirror: Information and Media in the Works of Oshii Mamoru and Kon Satoshi," *Canadian Journal of Film Studies* 18.1 (Spring 2009): 45.

12. Napier, *Anime from "Akira" to "Howl's Moving Castle,"* 149.

13. Sharalyn Orbaugh, "Busty Battlin' Babes: The Evolution of the Shōjo in 1990s Visual Culture," in *Gender and Power in the Japanese Visual Field*, ed. Joshua S. Mostow et al. (Honolulu: University of Hawai'i Press, 2003), 206. For an excellent introduction to *Shōjo* Studies, see Deborah Shamoon, *Passionate Friendship: The Aesthetics of Girls' Culture in Japan* (Honolulu: University of Hawai'i Press, 2012).

14. Orbaugh, "Busty Battlin' Babes," 225.

15. Yuko Hasegawa, "Post-Identity *Kawaii*: Commerce, Gender and Contemporary Japanese Art," in *Consuming Bodies: Sex and Contemporary Japanese Art*, ed. Fran Lloyd (London: Reaktion Books, 2002), 127.

16. While the English word "girl" can be applied to women of any age, *shōjo* is a more age-specific designation.

17. Monnet, "Towards the Feminine Sublime," 257.

18. Monnet, "Towards the Feminine Sublime," 258.

19. Monnet, "Towards the Feminine Sublime," 258.

20. *Paprika*, DVD, directed by Satoshi Kon (2006; Culver City, CA: Sony Pictures Home Entertainment, 2007).

21. Monnet, "Towards the Feminine Sublime," 248.

22. Thomas Lamarre, *The Anime Machine: A Media Theory of Animation* (Minneapolis: University of Minnesota Press, 2009), 85.

23. Alan Cholodenko, introduction to *The Illusion of Life 2: More Essays on Animation*, ed. Alan Cholodenko (Sydney: Power, 2007), 20.

24. Saitō, *Beautiful Fighting Girl*, 86.

25. Napier, *Anime from "Akira" to "Howl's Moving Castle,"* 22–23.

26. Hiroki Azuma, *Otaku: Japan's Database Animals* (Minneapolis: University of Minnesota Press, 2009), 10.

27. Susan Pointon, "Transcultural Orgasm as Apocalypse: *Urotsukidoji: The Legend of the Overfiend," Wide Angle* 19.3 (July 1997): 44.

28. Steven T. Brown, "Screening Anime," in *Cinema Anime*, ed. Stephen T. Brown (New York: Palgrave Macmillan, 2008), 6. While this might explain Huyghe and Parreno's passing interest in anime and manga, it does not justify the stories that they spin about rescuing Annlee from the hands of her Japanese captors.

29. Marshall McLuhan, *Understanding Media: The Extensions of Man* (New York: McGraw-Hill, 1965), 31.

30. Timothy Perper and Martha Cornog, "Psychoanalytic Cyberpunk Midsummer-Night's Dreamtime: Kon Satoshi's *Paprika*," *Mechademia* 4 (2009): 327.

31. Quoted in Philip Nobel, "Sign of the Times," *Artforum*, January 2003, Artforum.com, accessed July 24, 2013, artforum.com/inprint/issue=200301&id=3958&pagenum=1.

32. "Splitting of the Breast," *Neon Genesis Evangelion*, Episode 16, TV Tokyo (Tokyo, January 17, 1996).

33. Azuma, *Otaku*, 53. The author later clarifies that the character is really just the interface between the consumer and a database of character attributes, such as personality traits, hair and eye color, and costume elements, that are recycled and recombined.

34. Satoshi Kon, audio commentary, *Paprika* DVD.

35. Thomas Lamarre, "The First Time as Farce: Digital Animation and the Repetition of Cinema," in *Cinema Anime*, ed. Steven T. Brown (New York: Palgrave Macmillan, 2008), 179.

36. Lamarre, "The First Time as Farce," 182.

37. *Metropolis*, DVD, directed by Rintarō (2001; Culver City, CA: Sony Pictures Home Entertainment, 2002).

38. Lamarre, "The First Time as Farce," 182.

39. Bolter and Grusin, *Remediation: Understanding New Media* (Cambridge, MA: MIT Press, 1999), 273.

40. Bolter and Grusin, *Remediation*, 47.

41. Azuma, *Otaku*, 101.

42. Knight, "Return to the Movies, Return to the World: *Ratatouille* and *Paprika*," *Slant*, July 8, 2007, www.slantmagazine.com/house/2007/07/return-to-the-movies-return-to-the-world-ratatouille-paprika/.

43. Friedrich A. Kittler, *Gramophone, Film, Typewriter* (Stanford, CA: Stanford University Press, 1999), 1.

44. Henry Jenkins, *Convergence Culture: Where Old and New Media Collide* (New York: New York University Press, 2006), 26.

45. Jenkins, *Convergence Culture*, 2.

46. Ithiel de Sola Pool, *Technologies of Freedom* (Cambridge, MA: Belknap Press of Harvard University Press, 1983): 23.

47. Although Bolter and Grusin assert that "convergence is remediation under another name," they wrote that statement before Jenkins published what has since become the definitive explanation of media convergence within cinema and media studies. Bolter and Grusin, *Remediation*, 224.

48. Arild Fetveit, "Convergence by Means of Globalized Remediation," *Northern Lights* 5 (2007): 62.

49. Fetveit, "Convergence by Means of Globalized Remediation," 62.

50. Lev Manovich, "Alan Kay's Universal Media Machine," *Northern Lights* 5 (2007): 44.

51. Price, "IMG MGMT: Teen Image," October 22, 2009, www.artfagcity.com/2009/10/22/img-mgmt-teen-image/.

52. Knight, "Return to the Movies, Return to the World."

53. Monnet, "Towards the Feminine Sublime," 257.

54. Baudrillard, "The Ecstasy of Communication," in *Postmodern Culture*, ed. Hal Foster (London: Pluto, 1985), 132.

55. Baudrillard, "The Ecstasy of Communication," 131, 132.

56. Baudrillard, "The Ecstasy of Communication," 132, 131, 130, 131, 131.

57. Baudrillard, "The Ecstasy of Communication," 127. The Japanese intellectual movement "New Academism," which held wide appeal both inside and outside the Academy in the 1980s, was profoundly influenced by postmodern and poststructuralist thinkers such as Baudrillard and Jacques Derrida. See Jonathan E. Abel and Shion Kono, introduction to Azuma, *Otaku*, xx–xi.

58. "About Us," Metamorphose Temps de Fille, accessed July 29, 2013, www.metamorphose.gr.jp/company.html.

59. Anne Allison, *Millennial Monsters: Japanese Toys and the Global Imagination* (Berkeley: University of California Press, 2006), 10.

60. Saitō, *Beautiful Fighting Girl*, 160.

61. Saitō, *Beautiful Fighting Girl*, 160, 161.

62. Saitō, *Beautiful Fighting Girl*, 38.

63. Allison, *Millennial Monsters*, 140. Here, Allison is glossing Sharon Kinsella.

64. Azuma, *Otaku*, 53. Also, "Convergence does not occur through media appliances, however sophisticated they may become. Convergence occurs within the brains of individual consumers and through their social interactions with others." Jenkins, *Convergence Culture*, 3.

65. Saitō, *Beautiful Fighting Girl*, 86.

66. Vivian Sobchack, "Animation and Automation, or, The Incredible Effortfulness of Being," *Screen* 50.4 (Winter 2009): 375.

67. Marc Steinberg, "Characterizing a New Seriality: Murakami Takashi's DOB Project," *Parachute* 110 (2003): 100–101.

68. Spyros Papapetros, *On the Animation of the Inorganic: Art, Architecture, and the Extension of Life* (Chicago: University of Chicago Press, 2012), vii.

69. As Thomas Lamarre explains Takashi Murakami's strategic juvenilization of otaku, "Styling male otaku as 'little boys' is his way of signaling their ambivalent condition and their ambivalent response to military power. While constrained to remain boys under the security umbrella of 'manly' American military-industrial power, little boy otaku found solace in a two-fold attachment, at once to childhood and the objects associated with it (toys, anime, manga, games, dolls) and to the power of military technology." Lamarre, *The Anime Machine*, 117.

70. Perper and Cornog, "Psychoanalytic Cyberpunk Midsummer-Night's Dreamtime," 327.

71. In an unforgettably haptic and horrific rape scene evoking the tentacle pornography of *Urotsukidōji: Legend of the Overfiend* (1987–89), the Chairman's evil assistant jams his hand into Paprika's body and shoves it upward, splitting her in two from crotch to sternum. Underneath her skin and clothing, now a discarded, rubbery membrane, is Dr. Chiba, the rapist's intended victim. This seems to challenge my understanding of the film's spatial metaphysics of girlishness (the idea that the girl has an inside but no interior). However, any reading of the scene as a more brutal version of the stock unmasking of a superhero is undercut by the fact that the Chairman's henchman is currently in control of the nightmare. Driven mad with unrequited lust for Dr. Chiba, he fills Paprika's vacuity with the fullness of his own dream-woman and removes Paprika as an obstacle to his perverse plans. Paprika is still empty; she has just been filled with another version of herself.

72. Saitō, *Beautiful Fighting Girl*, 25.

73. *Alice in Wonderland: The Masterpiece Edition*, DVD, directed by Clyde Geronimi, Wilfred Jackson, and Hamilton Luske (1951; Burbank, CA, Walt Disney Home Video, 2004).

74. Teri Silvio, "Animation: The New Performance," *Journal of Linguistic Anthropology* 20.2 (2010): 422–38, 431.

75. Silvio, "Animation," 431.

76. Silvio, "Animation," 432.

77. "'Hardball with Chris Matthews' for Friday, September 14, 2012" (transcript), NBC News, September 17, 2012, www.msnbc.msn.com/id/49061711/ns/msnbc-hardball_with_chris_matthews/t/hardball-chris-matthews-friday-september-th/#.UJH6QRi5JEs.

78. USA Today, "USA Today Redesign: A Logo That Moves," YouTube video, :34, September 13, 2012, www.youtube.com/watch?v=cMD7GeYptos.

79. USA Today, "USA Today Redesign."

80. "'Hardball with Chris Matthews' for Friday, September 14, 2012."

81. "Apple 1984 Super Bowl Commercial Introducing Macintosh Computer (HD)," YouTube video, 1:03, June 25, 2013, www.youtube.com/watch?v=2zfqw8nhUwA.

82. Significantly, Apple had to destroy this screen. In 1983, the company could not yet conceive of the subtle and complex entwining of twenty-first-century media. It took some time for Apple to determine how it would keep television and film alive in order to use the mediums to its advantage. Their answer: don't kill these screens; shrink them.

83. John Whittier Treat, "Yoshimoto Banana Writes Home: The Shōjo in Japanese Popular Culture," in *Contemporary Japan and Popular Culture*, ed. John Whittier Treat (Richmond, UK: Curzon, 1996); G. Stanley Hall, *Educational Problems*, vol. 2 (New York: D. Appleton and Company, 1911), 7.

84. Frenchy Lunning, "Under the Ruffles: Shōjo and the Morphology of Power," *Mechademia* 6 (2011): 4.

AFTERWORD

1. Lewis Carroll, *Alice's Adventures in Wonderland* (Mineola, NY: Dover, 1993), 27–28.

2. Georges Didi-Huberman, "Knowledge: Movement (The Man Who Spoke to Butterflies)," foreword to *Aby Warburg and the Image in Motion*, by Philippe-Alain Michaud (Brooklyn: Zone Books, 2004), 14.

3. Carl Zimmer, "Nonfiction: Nabokov Theory on Butterfly Evolution is Vindicated," *New York Times*, January 25, 2011, www.nytimes.com/2011/02/01/science/01butterfly.html?pagewanted=all&_r=0.

4. G. Stanley Hall, *Educational Problems*, vol. 2 (New York: D. Appleton and Company, 1911), 1.

5. Scott Cutler Shershow, *Puppets and "Popular" Culture* (Ithaca, NY: Cornell University Press, 1995), 69.

6. Quoted in E. H. Gombrich, *Aby Warburg: An Intellectual Biography* (London: University of London Press, 1970), 110.

7. Albert Ploeger, "Mary and Anlee: About Representation and Presence in Visual Arts," in *Visual Arts and Religion*, ed. Hans Alma, Marcel Barnard, and Volker Küster (Berlin: Verlag, 2009), 31.

8. Here I am using Michel de Certeau's terminology from *The Practice of Everyday Life*.

9. Roy Ascott, "The Ambiguity of Self: Living in a Variable Reality," 2009, 1, accessed July 29, 2013, www.thenetworkedself.net/index.php/participants/roy-ascott.

10. Ascott, "The Ambiguity of Self," 3, 3, 6.

11. Ascott, "The Ambiguity of Self," 2.

12. Ascott, "The Ambiguity of Self," 4. Ascott expands upon an understanding of papillonage from utopian thinker Charles Fourier, who insists that variety is not the spice but the meat of life; the butterfly instinct is the need for variability of occupation.

13. That is, unless we are using Walter Benjamin's concept of the mimetic faculty, which is always generative.

14. Ascott, "The Ambiguity of Self," 2.

15. Quoted in Manovich, "Alan Kay's Universal Media Machine," *Northern Lights* 5 (2007): 15.

16. Anne Allison, *Millennial Monsters: Japanese Toys and the Global Imagination* (Berkeley: University of California Press, 2006), 134.

17. "Yes, We're Open Info," Netherlands Media Art Institute, accessed July 30, 2013, nimk.nl/eng/yes-were-open-info.

18. Gast (Tasha (@PimptressObject)), "Come Inside Girls," ANNLEE@NIMK.NL, May 24, 2012, nimk.nl/blog/annlee/2012/05/24/come-inside-girls/.

19. "Emilie Gervais and Sarah Weis," *Triangulation* Blog, August 16, 2012, www.triangu lationblog.com/2012/08/emilie-gervais-sarah-weis.html.

20. "Leisure Suit Larry Land of the Lounge Lizards-Trailer (1987)," YouTube video, 1:05, September 12, 2007, www.youtube.com/watch?v=XqXpWJhsuaI&feature=youtu.be.

21. The project also incorporates Twitter conversations and manga and anime-style figures as well as GIFs, photographs, and video stills of girls' faces contributed by users (at least, so it seems).

22. "Emilie Gervais and Sarah Weis."

23. Quoted in Ben Valentine, "Twitter as Art," *Hyperallergenic*, July 13, 2012, hyperallergic .com/54123/twitter-as-art/.

24. Quoted in Valentine, "Twitter as Art."

25. "Emilie Gervais and Sarah Weis."

26. "Emilie Gervais and Sarah Weis."

27. Another Annlee reboot is Tino Sehgal's performance *Ann Lee* (2011), in which several adolescent girls in succession asked visitors philosophical questions and ruefully reflected on their condition ("She lives passed from artist to artist," as Claire Bishop remembers Annlee's pitter-patter, "all of whom seem too busy to hang out with her"). See Claire Bishop, "Rooms for Improvement," *Artforum* diary, July 18, 2011, artforum.com/diary/id=28633.

28. CADIE, April 1, 2009 (6:13 p.m.), "goodbye," *CADIE's Personal World Wide Website*, accessed July 26, 2013, cadiesingularity.blogspot.com/2009/04/goodbye.html. I discuss this site in chapter 3.

29. CADIE, "goodbye."

30. This is an anonymously produced text attributed to Annlee and created in response to the Netherlands Media Art Institute's call for new works featuring the Annlee character. Gast (guest poster), "'Tis I, Veiled in Sin," July 2, 2012, nimk.nl/blog/annlee

/2012/07/02/tis-i-veiled-sin/. The title of the piece is hyperlinked to Annlee's Twitter page, which documents a Twitter performance that occurred between June 10 and July 27, 2012.

31. Haislip, "Sonnet for the Enom Domain Parking Girl," MichaelHaislip.com, January 23, 2008, web.archive.org/web/20080716070508/www.michaelhaislip.com/sonnet-for-the-enom-domain-parking-girl/. Enom is a company that provides web hosting and domain name registration. It uses iStockphoto's "Attractive Student" photograph on its parked domains.

32. Arild Fetveit, "Convergence by Means of Globalized Remediation," *Northern Lights* 5 (2007): 59. Here he is also critiquing Mark Hansen's *New Philosophy for New Media* (Cambridge, MA: MIT Press, 2006).

bibliography

"1960s Kodak Ad 'Turn Around.'" YouTube video, 2:00. April 30, 2007. www.youtube.com/watch?v=qBWVWjdNWCo.

Agamben, Giorgio. "Nymphs." In *Releasing the Image: From Literature to New Media*, edited by Jacques Kalip and Robert Mitchell, 60–80. Stanford, CA: Stanford University Press, 2011.

Allison, Anne. *Millennial Monsters: Japanese Toys and the Global Imagination*. Berkeley: University of California Press, 2006.

Anderson, Robert Helms, Philip M. Feldman, Scott Gerwehr, Brian Houghton, Richard Mesic, John Pinder, Jeff Rothenberg, and James Chiesa. *Securing the U.S. Defense Information Infrastructure: A Proposed Approach*. Santa Monica, CA: Rand, 1999. Accessed July 26, 2013. www.rand.org/pubs/monograph_reports/MR993.html.

Anno, Hideaki, and Hiroshi Yamaguchi. "Splitting of the Breast." *Neon Genesis Evangelion*, episode 16, directed by Kazuya Tsurumaki, aired January 17, 1996.

"Apple 1984 Super Bowl Commercial Introducing Macintosh Computer (HD)." YouTube video, 1:03. June 25, 2013. www.youtube.com/watch?v=2zfqw8nhUwA.

Ascott, Roy. "The Ambiguity of Self: Living in a Variable Reality." 2009. Available at www.thenetworkedself.net/index.php/participants/roy-ascott.

Azuma, Hiroki. *Otaku: Japan's Database Animals*. Minneapolis: University of Minnesota Press, 2009.

Barikin, Amelia. *Parallel Presents: The Art of Pierre Huyghe*. Cambridge, MA: MIT Press, 2012.

Baudrillard, Jean. "The Ecstasy of Communication." In *Postmodern Culture*, edited by Hal Foster, 126–34. London: Pluto, 1985.

Bauman, Zygmunt. *Liquid Modernity*. Cambridge, UK: Polity, 2000.

Bell, Elizabeth. "Somatexts at the Disney Shop: Constructing the Pentimentos of Women's Animated Bodies." In *From Mouse to Mermaid: The Politics of Film, Gender, and Culture*, edited by Elizabeth Bell, Lynda Haas, and Laura Sells, 107–24. Bloomington: Indiana University Press, 1995.

Bellmer, Hans. *Little Anatomy of the Physical Unconscious, or The Anatomy of the Image*. Translated by Jon Graham. Waterbury Center, Vt.: Dominion, 2004.

Berg, Louis. "Alice in Movie Land." *Los Angeles Times*, January 14, 1951.

Bishop, Claire. "Rooms for Improvement." *Artforum* diary, July 18, 2011. http://artforum.com/diary/id=28633.

Bolter, Jay David, and Richard Grusin. *Remediation: Understanding New Media*. Cambridge, MA: MIT Press, 1999.

Bonnenfant, Katya. "Towards the Anonymity of the Artist." Accessed July 26, 2013. www.katya-bonnenfant.com/annesophie/comments04.html.

Bourriaud, Nicolas. *Postproduction*. New York: Lukas and Sternberg, 2005. http://ebookbrowse.com/bourriaud-postproduction-pdf-d49348259.

———. "Relational Aesthetics." In *Participation*, edited by Claire Bishop, 160–71. London: Whitechapel, 2006.

———. "The Reversibility of the Real: Nicolas Bourriaud on Pierre Huyghe." *Tate, etc.* 7, Summer 2006. www.tate.org.uk/tateetc/issue7/pierrehuyghe.htm.

Brown, Steven T. "Screening Anime." In *Cinema Anime*, edited by Steven T. Brown, 1–19. New York: Palgrave Macmillan, 2008.

"Bunny Greenhouse." YouTube video. Last modified January 6, 2009. www.youtube.com/user/BunnyGreenhouse.

Burnett, Robert, and P. David Marshall. *Web Theory: An Introduction*. Abingdon: Routledge, 2003.

Burnett, Ron. "From Photography to Imography: New Media as Metaphor." In *Fluid Screens, Expanded Cinema*, edited by Janine Marchessault and Susan Lord, 126–44. Toronto: University of Toronto Press, 2007.

CADIE's Personal World Wide Website. Accessed July 26, 2013. http://cadiesingularity.blogspot.com/2009/04/goodbye.html.

Campanelli, Vito. *Web Aesthetics*. Rotterdam: NAi, 2010.

Carnegie Mellon University. "Alice." Video, 3:59. Accessed October 30, 2012. www.alice.org/index.php?page=what_is_alice/what_is_alice.

Carroll, Lewis. *Alice's Adventures in Wonderland*. Mineola, NY: Dover, 1993.

Carter, Huntly. *The New Spirit in the Cinema*. London: Harold Shaylor, 1930.

Carter, Steven. "Real Simulacra Redux: Barbie and Jane vs. the Wooden Nutmegs of Connecticut." *Journal of Popular Culture* 27.3 (Winter 1993): 41–59.

Castañeda, Claudia. *Figurations: Child, Bodies, Worlds*. Durham, NC: Duke University Press, 2002.

Catmull, Ed, and Frank Parke. "A Computer Animated Hand." Vimeo video, 6:32. Posted by "Robby Ingebretsen." 2011. vimeo.com/16292363.

Cholodenko, Alan. Introduction to *The Illusion of Life 2: More Essays on Animation*, edited by Alan Cholodenko, 13–95. Sydney: Power, 2007.

Chun, Wendy Hui Kyong. "The Enduring Ephemeral, or the Future Is a Memory." *Critical Inquiry* 35.1 (Autumn 2008): 148–71.

Clark, David. "The Discrete Charm of the Digital Image: Animation and New Media." In *The Sharpest Point: Animation at the End of Cinema*, edited by Chris Gehman and Steve Reinke, 138–51. Toronto: YYZ Books, 2005.

Cloninger, Curt. "Mouchette Has a Posse." *Rhizome*, March 5, 2001. rhizome.org/artbase/artwork/2261/.

Cohen, Morton N., ed. *Lewis Carroll and the Kitchens*. New York: Lewis Carroll Society of North America, 1980.

"Come Inside Girls." ANNLEE@NIMK.NL, May 24, 2012. http://nimk.nl/blog/annlee/2012/05/24/come-inside-girls/.

Cotter, Holland. "Art Review: Films That Keep Asking, Is It Fact or Fiction?" *New York Times*, January 19, 2001. www.nytimes.com/2001/01/19/movies/art-review-films-that-keep-asking-is-it-fact-or-fiction.html?pagewanted=2&src=pm.

Crafton, Donald. *Shadow of a Mouse: Performance, Belief, and World-Making in Animation*. Berkeley: University of California Press, 2013.

Danto, Arthur C. "Photography and Performance: Cindy Sherman's Stills." In *Cindy Sherman: Untitled Film Stills*, 5–14. New York: Rizzoli, 1990.

de Certeau, Michel. *The Practice of Everyday Life*. Berkeley: University of California Press, 1988.

Deleuze, Gilles. *Cinema 2: The Time Image*. Translated by Hugh Tomlinson and Robert Galeta. Minneapolis: University of Minnesota Press, 2001.

Deleuze, Gilles, and Félix Guattari. *A Thousand Plateaus: Capitalism and Schizophrenia*. Minneapolis: University of Minnesota Press, 1987.

Didi-Huberman, Georges. "Knowledge: Movement (The Man Who Spoke to Butterflies)." Foreword to *Aby Warburg and the Image in Motion*, by Philippe-Alain Michaud, 7–19. Brooklyn: Zone Books, 2004.

Dietrich, Frank. "Visual Intelligence: The First Decade of Computer Art (1965–1975)." *Leonardo* 19.2 (1986): 159–69.

"Doll Maker." *Rozen Maiden: Träumend*. TBS. Tokyo, December 8, 2005.

Dowson, Mark, Humphrey Evans, IBM, Andrew Logan, and Peter Schmidt. "Glossary." In *Cybernetic Serendipity: Computer and the Arts*, edited by Jasia Reichardt, 14–16. London: Studio International, 1968.

Driscoll, Catherine. *Girls: Feminine Adolescence in Popular Culture and Cultural Theory*. New York: Columbia University Press, 2002.

Eisenstein, Sergei. *Eisenstein on Disney*. Edited by Jay Leyda. London: Methuen, 1988.

Elfving, Taru. "Haunted: Writing with the Girl." In *Girls! Girls! Girls! in Contemporary Art*, edited by Catherine Grant and Lori Waxman, 107–24. Bristol: Intellect, 2011.

"Emilie Gervais and Sarah Weis." Triangulation (blog), August 16, 2012. www.triangulationblog.com/2012/08/emilie-gervais-sarah-weis.html.

Escalas, Jennifer Edson, Kapil Jain, and Judi E. Strebel. "Satisfaction, Frustration, and Delight: A Framework for Understanding How Consumers Interact with Web Sites." In *Internet Marketing Research: Theory and Practice*, edited by Ook Lee, 231–51. Hershey, PA: Idea Group, 2001.

"Exhibitions: Cybernetic Serendipity." *Time*, October 4, 1968. www.time.com/time/magazine/article/0,9171,838821,00.html#ixzz2AotPKMB3.

Fetveit, Arild. "Convergence by Means of Globalized Remediation." *Northern Lights* 5 (2007): 57–74.

Foster, Hal. *Compulsive Beauty*. Cambridge, MA: MIT Press, 1997.

Franklin Furnace. "Mouchette." Accessed July 25, 2013. franklinfurnace.org/artists/funded_projects/artistsPages/mouchette/mouchette.html.

Gardner, William O. "The Cyber Sublime and the Virtual Mirror: Information and

Media in the Works of Oshii Mamoru and Kon Satoshi." *Canadian Journal of Film Studies* 18.1 (Spring 2009): 44–70.

Gervais, Emilie. *Parked Domain Girl Tombstone*. Accessed July 26, 2013. parkeddomain girltombstone.net.

Ghost in the Shell. DVD. Directed by Mamoru Oshii. 1996; Los Angeles: Manga Entertainment, 1998.

Giaccardi, Elisa. "Movements and Passages: The Legacy of Net Art." *Leonardo Electronic Almanac* 13.4 (April 2005): 23–32. www.leoalmanac.org/wp-content/uploads/2012/07/LEA-v13-n4.pdf.

Gillen, Marilyn A. "For Entertainment Biz, It Was Year of the Morph." *Billboard*, December 24, 1994.

Gombrich, E. H. *Aby Warburg: An Intellectual Biography*. London: University of London Press, 1970.

Goodall, Jane. "Transferred Agencies: Performance and the Fear of Automatism." *Theatre Journal* 49.4 (1997): 441–53.

Google. "Goodbye." YouTube video, :51. March 27, 2009. www.youtube.com/watch?v=BeJ9Q40kIR0.

Grau, Oliver, and Thomas Veigl. Introduction to *Imagery in the 21st Century*, edited by Oliver Grau and Thomas Veigl, 1–17. Cambridge, MA: MIT Press, 2011.

Greer, Germaine. "Mad about the Boy." *Guardian*, May 14, 2005. www.guardian.co.uk/film/2005/may/14/film.

Hadjioannou, Markos. *From Light to Byte: Toward an Ethics of Digital Cinema*. Minneapolis: University of Minnesota Press, 2012.

Haislip, Michael. "Sonnet for the Enom Domain Parking Girl." MichaelHaislip.com, January 23, 2008. web.archive.org/web/20080716070508/www.michaelhaislip.com/sonnet-for-the-enom-domain-parking-girl/.

Halberstam, Judith. "Queerreality." *Empyre* forum, June 12, 2012. www.mail-archive.com/empyre@lists.cofa.unsw.edu.au/msg04219.html.

Hall, G. Stanley. *Educational Problems* Vol. 2. New York: D. Appleton and Company, 1911.

Hansen, Mark B. N. *New Philosophy for New Media*. Cambridge, MA: MIT Press, 2006.

Harris, Anita. *Future Girl: Young Women in the Twenty-First Century*. New York: Routledge, 2004.

Harris, Daniel. *Cute, Quaint, Hungry and Romantic: The Aesthetics of Consumerism*. New York: Da Capo, 2001.

Hasegawa, Yuko. "Post-Identity *Kawaii*: Commerce, Gender and Contemporary Japanese Art." In *Consuming Bodies: Sex and Contemporary Japanese Art*, edited by Fran Lloyd, 127–41. London: Reaktion Books, 2002.

Hauptman, Jodi. *Joseph Cornell: Stargazing in the Cinema*. New Haven, CT: Yale University Press, 1999.

Holmes, Brian. "The Flexible Personality: For a New Cultural Critique." 2001. eipcp.net/transversal/1106/holmes/en.

Hutchinson, Jamie. "Culture, Communication, and an Information Age Madonna." *IEEE Professional Communication Society Newsletter* 45.3 (May/June 2001): 4–7.

Huyghe, Pierre. "Two Minutes Out of Time." In *No Ghost Just a Shell*, edited by Pierre Huyghe and Philippe Parreno, 44–57. Cologne: Verlag der Buchhandlung Walter König, 2003.

Huyghe, Pierre, Stefan Kalmar, Philippe Parreno, Beatrix Ruf, and Hans Ulrich Obrist. "Conversations." In *No Ghost Just a Shell*, edited by Pierre Huyghe and Philippe Parreno, 14–30. Cologne: Verlag der Buchhandlung Walter König, 2003.

iStockphoto. "Attractive Student." Accessed July 26, 2013. www.istockphoto.com/stock-photo-746781-attractive-student.php.

Ito, Parker. "The Most Infamous Girl in the History of the Internet / Attractive Student / Parked Domain Girl." Accessed July 26, 2013. www.parkerito.com/parker_ito/.

Jamieson, Helen Varley. "Could You Be the Next Mouchette?" *Rhizome*, April 14, 2003. rhizome.org/editorial/2003/apr/14/could-you-be-the-next-mouchette/.

Jenkins, Henry. *Convergence Culture: Where Old and New Media Collide*. New York: New York University Press, 2006.

———. "Digital Land Grab." *MIT Technology Review*, March 1, 2000. www.technologyreview.com/article/400696/digital-land-grab/.

Johannesen, Hanne-Louise. "Performing Identity." Accessed July 27, 2013. www.usynligebyer.dk/fra_staff/publikationer/performing_identity.html.

Johnson, Paddy. "Calling London! Free Ann Lee from the Tate." *Art Fag City*, October 12, 2009. www.artfagcity.com/2009/10/12/calling-london-free-ann-lee-from-the-tate/.

Johnson, Stephen. *Interface Culture: How New Technology Transforms the Way We Create and Communicate*. New York: Basic Books, 1997.

Joselit, David. *After Art*. Princeton, NJ: Princeton University Press, 2013.

———. *Feedback: Television against Democracy*. Cambridge, MA: MIT Press, 2007.

Kay, Alan, and Adele Goldberg. "Personal Dynamic Media." In *Multimedia: From Wagner to Virtual Reality*, edited by Randall Packer and Ken Jordan, 167–78. New York: W. W. Norton, 2001.

Kelly, Kevin. "Gossip Is Philosophy." *Wired*, May 1995. www.wired.com/wired/archive/3.05/eno_pr.html.

Kelsey, John. "John Kelsey on Tim Burton's Alice in Wonderland." *Artforum* (May 2010). www.thefreelibrary.com/Lost+in+Space%3A+JOHN+KELSEY+ON+TIM+BURTON%27S+ALICE+IN+WONDERLAND.-a0227296425.

Kirby, Alan. *Digimodernism: How New Technologies Dismantle the Postmodern and Reconfigure Our Culture*. New York: Continuum Books, 2009.

Kittler, Friedrich A. *Gramophone, Film, Typewriter*. Stanford, CA: Stanford University Press, 1999.

Knight, Ryland Walker. "Return to the Movies, Return to the World: *Ratatouille* and *Paprika*." *Slant*, July 8, 2007. www.slantmagazine.com/house/2007/07/return-to-the-movies-return-to-the-world-ratatouille-paprika/.

Knotenpunkte. "Mouchette: Museum of Contemporary Art, Siegen." Accessed July 25, 2013. knotenpunkte.net/kp-en/artist7.html.

Kracauer, Siegfried. "The Mass Ornament." In *The Mass Ornament: Weimar Essays*, 75–88. Cambridge, MA: Harvard University Press, 1995.

Kumar, Karthik, Huigao Duan, Ravi S. Hegde, Samuel C. W. Koh, Jennifer N. Wei, and

Joel K. W. Yang. "Printing Colour at the Optical Diffraction Limit." *Nature Nanotechnology* 7.9 (2012): 557–61.

Lamarre, Thomas. *The Anime Machine: A Media Theory of Animation*. Minneapolis: University of Minnesota Press, 2009.

———. "The First Time as Farce: Digital Animation and the Repetition of Cinema." In *Cinema Anime*, edited by Stephen T. Brown, 161–88. New York: Palgrave Macmillan, 2008.

Latham, Robert. *Consuming Youth: Vampires, Cyborgs, and the Culture of Consumption*. Chicago: University of Chicago Press, 2002.

"Leisure Suit Larry Land of the Lounge Lizards-Trailer (1987)." YouTube video, 1:05. September 12, 2007. www.youtube.com/watch?v=XqXpWJhsuaI&feature=youtu.be.

Lialina, Olia. "Early Experiments Online." *Guggenheim*, July 22, 2010. www.guggenheim.org/new-york/interact/participate/youtube-play/the-take/moving-images/3583-early-experiments-online.

Lichtenstein, Therese. *Behind Closed Doors: The Art of Hans Bellmer*. Berkeley: University of California Press, 2001.

Lichty, Patrick. "Just What Is It That Makes New Media So Different, So Appealing?" *NMC Media-N: Journal of the New Media Caucus* 5.1 (Spring 2009). median.newmediacaucus.org/archives_in_progress/pre_2009_issues/issues.php?f=papers&time=2009_spring&page=lichty.

Licklider, J. C. R., and Robert Taylor. "The Computer as a Communication Device." In *In Memorium: J. C. R. Licklider (1915–1990)*, 21–41. Palo Alto: Systems Research Center, 1990.

Lonergan, Guthrie. "Everything." Accessed July 27, 2013. www.theageofmammals.com/.

Longstaff, Thomas, James T. Ellis, Shawn V. Hernan, Howard F. Lipson, Robert D. McMillan, Linda Hutz Pesante, and Derek Simmel. "Security of the Internet." In *The Froehlich/Kent Encyclopedia of Telecommunications*, vol. 15, edited by Fritz E. Froehlich and Allen Kent, 231–56. New York: Marcel Dekker, 1998.

Lunenfeld, Peter, ed. *The Digital Dialectic: New Essays on New Media*. Cambridge, MA: MIT Press, 1999.

———. "Hall of Mirrors: Art and Film since 1945." *Artforum* 34.10 (Summer 2006). www.thefreelibrary.com/Hall+of+Mirrors%3A+Art+and+Film+since+1945.-a018533856.

Lunning, Frenchy. "Under the Ruffles: Shōjo and the Morphology of Power." *Mechademia* 6 (2011): 3–19.

Malabou, Catherine. *What Should We Do with Our Brain?* New York: Fordham University Press, 2008.

Manovich, Lev. "Alan Kay's Universal Media Machine." *Northern Lights* 5 (2007): 39–56.

———. "Art after Web 2.0." In *The Art of Participation: 1950–Now*, edited by Rudolf Frieling, 66–80. New York: Thames and Hudson, 2008.

———. "Database as Symbolic Form." In *Database Aesthetics: Art in the Age of Information Overload*, edited by Victoria Vesna, 39–60. Minneapolis: University of Minnesota Press, 2007.

———. *The Language of New Media*. Cambridge, MA: MIT Press, 2001.

Matrix, Sydney Eve. *Cyberpop: Digital Lifestyles and Commodity Culture*. New York: Routledge, 2006.

The Matrix, directed by Lana and Andy Wachowski. 1999. Warner Brothers Pictures.

Mayfield, Kendra. "Art Explores Cartoon as Commodity." *Wired*, December 14, 2002. www.wired.com/culture/lifestyle/news/2002/12/56762.

McCracken, Grant. "'Consumers' or 'Multipliers.'" *Spreadable Media*. Accessed July 27, 2013. http://spreadablemedia.org/essays/mccracken/#.UeL5deDvNuU.

———. *Transformations: Identity Construction in Contemporary Culture*. Bloomington: Indiana University Press, 2008.

McDonough, Tom. "No Ghost." *October* 110 (Fall 2004): 107–30.

McHugh, Gene. *Post Internet*. Accessed July 26, 2013. http://122909a.com/?p=2054.

McKay, Sally. *Digital Media Tree*. November 25, 2003. www.digitalmediatree.com/sally mckay/?24066.

McLuhan, Marshall. *The Mechanical Bride: Folklore of Industrial Man*. Boston: Beacon, 1968.

———. *Understanding Media: The Extensions of Man*. New York: McGraw-Hill, 1965.

Menning, Chris. "Who Is 'Parked Domain Girl' and Can I Marry Her?" Accessed July 26, 2013. www.urlesque.com/2010/11/03/parked-domain-girl/.

Merchant, Brian. "Lenna, the First Lady of the Internet." *Motherboard*, June 2012. motherboard.vice.com/read/meet-lenna-the-first-lady-of-the-internet--2.

Merritt, Russell, and J. B. Kaufman. *Walt in Wonderland: The Silent Films of Walt Disney*. Baltimore: Johns Hopkins University Press, 1993.

Metamorphose Temps de Fille. "About Us." Accessed July 29, 2013. www.metamorphose.gr.jp/company.html.

Metropolis. DVD. Directed by Rintarō. 2001; Culver City, CA: Sony Pictures Home Entertainment, 2002.

Microsoft. "How to Modify the Number of Undo Levels in Excel." Accessed July 22, 2013. support.microsoft.com/kb/211922.

Mitchell, W. J. *The Reconfigured Eye: Visual Truth in the Post-Photographic Era*. Cambridge, MA: MIT Press, 1992.

Mitchell, W. J. T. *What Do Pictures Want?: The Lives and Loves of Images*. Chicago: University of Chicago Press, 2005.

M/M Paris. "No Ghost Just a Shell." Accessed July 25, 2013. www.mmparis.com/noghost.html.

Monnet, Livia. "Towards the Feminine Sublime, or The Story of 'a Twinkling Monad, Shape-Shifting across Dimension': Intermediality, Fantasy and Special Effects in Cyberpunk Film and Animation." *Japan Forum* 14.2 (2002): 225–68.

Monteiro, Stephen. "Lovely Gardens and Dark Rooms: Alice, the Queen, and the Spaces of Photography." In *Alice beyond Wonderland: Essays for the 21st Century*, edited by Christopher Hollingsworth, 101–16. Iowa City: University of Iowa Press, 2009.

Mouchette. Accessed July 25, 2013. Mouchette.org.

Mouchette. Accessed July 27, 2013. drivedrive.com/mouchette/power_plant.html.

Mouchette. "Clickable Mouchette." Accessed July 26, 2013. lituraterre.free.fr/qcmouchette/click.html.

Mouchette with Manthos Santorineos. "Rape, Murder and Suicide Are Easier When You Use a Keyboard Shortcut: Mouchette, an On-Line Virtual Character." *LEONARDO* 38.3 (2005): 203–6.

"Mouchette Censored by Robert Bresson's Widow." Accessed July 26, 2013. drivedrive .com/mouchette/censored.html.

MTAA. "Net Art NYC April 20, 21, 2003." April 22, 2003. www.mtaa.net/mtaaRR/news /twhid/photos/jodi_mouchette.html.

Murphie, Andrew, and John Potts. *Culture and Technology*. New York: Palgrave Macmillan, 2003.

Nabokov, Vladimir. *Lolita*. New York: Random House, 1977.

NBC News. "'Hardball with Chris Matthews' for Friday, September 14, 2012" (transcript). September 17, 2012. www.msnbc.msn.com/id/49061711/ns/msnbc-hardball_with _chris_matthews/t/hardball-chris-matthews-friday-september-th/#.UJH6QRi5JEs.

Nagy, Jeff. "A Little Brooke of Visions." *Bomblog*, September 26, 2012. bombsite.com /issues/1000/articles/6764.

Nake, Frieder. "Notes on the Programming of Computer Graphics." In *Cybernetic Serendipity: Computer and the Arts*, edited by Jasia Reichardt, 77–78. London: Studio International, 1968.

Napier, Susan. *Anime from "Akira" to "Howl's Moving Castle": Experiencing Contemporary Japanese Animation*. New York: Palgrave Macmillan, 2005.

Neddam, Martine. "Zen and the Art of Database Maintenance." In *Archive 2020: Sustainable Archiving of Born-Digital Cultural Content*, edited by Annet Dekker, 10–19. Amsterdam: Virtueel Platform, 2010.

Neddam, Martine, and Annet Dekker. "In Search of the Unexpected: Martine Neddam in Conversation with Annet Dekker," 65–73. In *Navigating E-Culture*, edited by Cathy Brickwood and Annet Dekker. Amsterdam: Virtueel Platform, 2009.

Negarestani, Reza. "Death as Perversion: Openness as Germinal Death." *CTHEORY*, October 15, 2003. www.ctheory.net/articles.aspx?id=396.

Negroponte, Nicholas. "Repurposing the Material Girl." *Wired*, November 1993. www .wired.com/wired/archive/1.05/negroponte.html.

Netherlands Media Art Institute. "Yes, We're Open Info." Accessed July 30, 2013. nimk .nl/eng/yes-were-open-info.

Ngai, Sianne. "The Cuteness of the Avant-Garde." *Critical Inquiry* 31.4 (Summer 2005): 811–47.

———. "'A Foul Lump Started Making Promises in My Voice': Race, Affect, and the Animated Subject." *American Literature* 74.3 (2002): 571–601.

———. *Our Aesthetic Categories: Zany, Cute, Interesting*. Cambridge, MA: Harvard University Press, 2012.

"Nicole Appelman." *Tim Gunn's Guide to Style*, season 1, episode 3, aired September 20, 2007.

Nobel, Philip. "No Ghost Just a Shell: The Ann Lee Project." Artforum.com, September 1, 2002. artforum.com/archive/id=3198.

———. "Sign of the Times." *Artforum*, January 2003. artforum.com/inprint/issue=200301&id=3958&pagenum=1.

Orbaugh, Sharalyn. "Busty Battlin' Babes: The Evolution of the Shōjo in 1990s Visual Culture." In *Gender and Power in the Japanese Visual Field*, edited by Joshua S. Mostow, Norman Bryson, and Maribeth Graybill, 201–27. Honolulu: University of Hawai'i Press, 2003.

Papapetros, Spyros. *On the Animation of the Inorganic: Art, Architecture, and the Extension of Life*. Chicago: University of Chicago Press, 2012.

Paprika. Directed by Satoshi Kon. 2006; Culver City, CA: Sony Pictures Home Entertainment, 2007.

"Parked Domain Girl Painting." *Buzzfeed*. Accessed July 26, 2013. www.buzzfeed.com/peggy/parked-domain-girl-painting.

Parreno, Philippe. "Anywhere out of the World." In *No Ghost Just a Shell*, edited by Pierre Huyghe and Philippe Parreno, 32–43. Cologne: Verlag der Buchhandlung Walter König, 2003.

Perper, Timothy, and Martha Cornog. "Psychoanalytic Cyberpunk Midsummer-Night's Dreamtime: Kon Satoshi's *Paprika*." *Mechademia* 4 (2009): 326–29.

Pilinovsky, Helen. "Body as Wonderland: Alice's Graphic Iteration in *Lost Girls*." In *Alice beyond Wonderland: Essays for the 21st Century*, edited by Christopher Hollingsworth, 175–98. Iowa City: University of Iowa Press, 2009.

"Plastic." Merriam-Webster. Accessed October 30, 2012. www.merriam-webster.com/dictionary/plastic.

Ploeger, Albert. "Mary and Anlee: About Representation and Presence in Visual Arts." In *Visual Arts and Religion*, edited by Hans Alma, Marcel Barnard, and Volker Küster, 25–38. Berlin: Verlag, 2009.

Pointon, Susan. "Transcultural Orgasm as Apocalypse: *Urotsukidoji: The Legend of the Overfiend*." *Wide Angle* 19.3 (July 1997): 41–63.

Pool, Ithiel de Sola. *Technologies of Freedom*. Cambridge, MA: Belknap Press of Harvard University Press, 1983.

Power, Nina. "She's Just Not That Into You." *Radical Philosophy* 177 (January/February 2013). www.radicalphilosophy.com/web/rp177-shes-just-not-that-into-you.

President's Information Technology Advisory Committee. *Cyber Security: A Crisis of Prioritization*. Arlington, VA: National Coordination Office for Information Technology Research and Development, 1995. Accessed July 26, 2013. www.nitrd.gov/Publications/PublicationDetail.aspx?pubid=45.

Price, Seth. "Dispersion." 2002. Accessed December 4, 2010. www.distributedhistory.com/Disperzone.html.

———. "IMG MGMT: Teen Image." *Art Fag City*, October 22, 2009. www.artfagcity.com/2009/10/22/img-mgmt-teen-image/.

Reichardt, Jasia. "Computer Art." In *Cybernetic Serendipity: Computer and the Arts*, edited by Jasia Reichardt, 70–71. London: Studio International, 1968.

Saitō, Tamaki. *Beautiful Fighting Girl*. Minneapolis: University of Minnesota Press, 2011.

Scanlan, Joe, and Kristel Van Audenaeren. "Roll Back the Stone: AnnLee Interviews Joe Scanlan." Accessed July 25, 2013. www.thingsthatfall.com/annlee.php.

Schneider, Dan. "Tale and Performance." *Culture Wars*, March 18, 2010. www.culturewars.org.uk/index.php/site/article/tale_and_performance/.

Shamoon, Deborah. *Passionate Friendship: The Aesthetics of Girls' Culture in Japan*. Honolulu: University of Hawai'i Press, 2012.

Shershow, Scott Cutler. *Puppets and "Popular" Culture*. Ithaca, NY: Cornell University Press, 1995.

Shinko, Kathryn. *Figaro Girl*, August 2010. figarogirl.wordpress.com/2010/08/.

Silvio, Teri. "Animation: The New Performance." *Journal of Linguistic Anthropology* 20.2 (2010): 422–38.

SKOR Foundation for Art and Public Domain. "How to Be Pink and Conceptual at the Same Time: A Conversation between Martine Neddam and Annet Dekker, Amsterdam, May 2011." Accessed July 25, 2013. www.skor.nl/eng/projects/item/interview-martine-neddam.

Sobchack, Vivian. "Animation and Automation, or, The Incredible Effortfulness of Being." *Screen* 50.4 (Winter 2009): 375–91.

———. "Introduction." In *Meta-morphing: Visual Transformation and the Culture of Quick-Change*, edited by Vivian Sobchack, xi–xxiii. Minneapolis: University of Minnesota Press, 2000.

Sollfrank, Cornelia. "Not Every Hacker Is a Woman." *Hacker Kulture*. Accessed July 27, 2013. www.dvara.net/HK/hackwoman.asp.

Solomon, Charles. "Back to Disney Studio: Adventures of Alice and Wendy." *Los Angeles Times*, January 3, 1983.

Solomon-Godeau, Abigail. "Suitable for Framing: The Critical Recasting of Cindy Sherman (1991)." In *Cindy Sherman*, edited by Johanna Burton, 53–64. Cambridge, MA: MIT Press, 2006.

"Sprint Girl." YouTube video, 0:31. September 4, 2012. www.youtube.com/watch?feature=player_embedded&v=aQnsN6eXbR8.

Spuybroek, Lars. *NOX: Machining Architecture*. New York: Thames and Hudson, 2004.

Stallabrass, Julian. *Internet Art: The Online Clash of Culture and Commerce*. London: Tate, 2003.

Stefans, Brian. "Third Hand Plays: The Comedy of Simulation." San Francisco Museum of Modern Art, August 16, 2011. blog.sfmoma.org/2011/08/third-hand-plays-the-comedy-of-simulation/.

Steinberg, Marc. "Characterizing a New Seriality: Murakami Takashi's DOB Project." *Parachute* 110 (2003): 90–109.

Stevenson, Seth. "I'd Like to Buy the World a Shelf-Stable Children's Lactic Drink." *New York Times Magazine*, March 10, 2002. www.nytimes.com/2002/03/10/magazine/i-d-like-to-buy-the-world-a-shelf-stable-children-s-lactic-drink.html?pagewanted=all&src=pm.

Stiegler, Bernard. *Taking Care of Youth and the Generations*. Translated by Stephen Barker. Stanford, CA: Stanford University Press, 2010.

Streeter, Thomas. *The Net Effect: Romanticism, Capitalism, and the Internet.* New York: New York University Press, 2011.

Sutherland, Ivan. "The Ultimate Display." In *Multimedia: From Wagner to Virtual Reality*, edited by Randall Packer and Ken Jordan, 232–36. New York: W. W. Norton, 2001.

Sweeney, Kathleen. *Maiden USA*. New York: Peter Lang, 2008.

Szita, Jane. "Conference Report." *Leonardo Reviews*, May 2001. www.leonardo.info/reviews/may2001/ev_CODE_szita.html.

"La Tana di Alice." Accessed October 30, 2012. www.pinocchio.it/old/uk/eeuropa.htm.

Tanner, Marcia. "Requiem for a Mail Order Bride." *Stretcher: Visual Culture in the San Francisco Bay Area and Beyond*. Accessed July 24, 2013. www.stretcher.org/features/no_ghost_just_a_shell/.

Tate. "No Ghost Just a Shell." Accessed July 28, 2010. www.tate.org.uk/servlet/CollectionDisplays?venueid=2&roomid=5676.

"Things Going On: From Play to Discover." Accessed October 20, 2012. www.pinocchio.it/old/uk/luoghi.htm.

Thomas, Frank, and Ollie Johnston. *The Illusion of Life: Disney Animation.* New York: Hyperion, 1981.

Time. "The New Face of America: Nov. 18, 1993." Accessed July 24, 2013. www.time.com/time/covers/0,16641,19931118,00.html.

Tiqqun. *Preliminary Materials for a Theory of the Young-Girl.* Translated by Ariana Reines. Los Angeles: Semiotext(e), 2012.

———. *Raw Materials for a Theory of the Young-Girl.* Anonymous translator and publisher. 2001. Accessed July 25, 2013. younggirl.jottit.com/.

"Tis I, Veiled in Sin." ANNLEE@NIMK.NL, July 2, 2012. nimk.nl/blog/annlee/2012/07/02/tis-i-veiled-sin/.

Treat, John Whittier. "Yoshimoto Banana Writes Home: The Shōjo in Japanese Popular Culture." In *Contemporary Japan and Popular Culture*, edited by John Whittier Treat, 275–308. Richmond, UK: Curzon, 1996.

Tribe, Mark, and Reena Jana. *New Media Art*. Cologne: Taschen, 2006.

USA Today. "USA Today Redesign: A Logo That Moves." YouTube video, :34. September 13, 2012. www.youtube.com/watch?v=cMD7GeYptos.

Valentine, Ben. "Twitter as Art." *Hyperallergenic*, July 13, 2012. hyperallergic.com/54123/twitter-as-art/.

Vélez, Pedro. "Hell in Lamb UC." Single page within *Hell in Lamb UC*. Accessed March 25, 2009. www.myspace.com/annleelives.

———. *Hell in Lamb UC*. Accessed March 25, 2009. www.myspace.com/hellinlambuc.

Verwoert, Jan. "Copyright, Ghosts and Commodity Fetishism." In *No Ghost Just a Shell*, edited by Pierre Huyghe and Philippe Parreno, 184–92. Cologne: Verlag der Buchhandlung Walter König, 2003.

Virilio, Paul. "The Aesthetics of Disappearance." In *The Paul Virilio Reader*, edited by Steve Redhead, 57–82. New York: Columbia University Press, 2004.

Volkart, Yvonne. "Connective Identities." Accessed July 26, 2013. www.obn.org/reading_room/writings/html/connective_id.html.

Waxman, Lori. "Through the Looking Glass with Heart-Shaped Sunglasses: Searching for Alice and Lolita in Contemporary Representations of Girls." In *Girls! Girls! Girls! in Contemporary Art*, edited by Catherine Grant and Lori Waxman, 17–44. Bristol: Intellect, 2011.

Weinbaum, Alys Eve, Lynn M. Thomas, Priti Ramamurthy, Uta G. Poiger, Madeleine Yu Dong, and Tani E. Barlow, ed. *The Modern Girl around the World: Consumption, Modernity, and Globalization*. Durham, NC: Duke University Press, 2008.

Wells, Paul. *Animation: Genre and Authorship*. London: Wallflower, 2002.

"What Is Alice?" Accessed October 30, 2012. www.alice.org/index.php?page=what_is_alice/what_is_alice.

Wikipedia. "List of Highest-Grossing Films." Last modified October 30, 2012. en.wikipedia.org/wiki/List_of_highest-grossing_films.

———. "Tears in Rain Soliloquy." Last modified July 26, 2013. en.wikipedia.org/wiki/Tears_in_rain_soliloquy.

Witchard, Anne. "Chinoiserie Wonderlands of the Fin de Siècle: Twinkletoes in Chinatown." In *Alice beyond Wonderland: Essays for the 21st Century*, edited by Christopher Hollingsworth, 155–73. Iowa City: University of Iowa Press, 2009.

Wolff, Rachel M. "We Bought a Virgin: The Issue of the Artist in *No Ghost Just a Shell*." *Shift: Graduate Journal of Visual and Material Culture* 4 (2011). shiftjournal.org/archives/articles/2011/wolff.pdf.

Yap, Chin-Chin. "Escaping Ghost." *ArtAsiaPacific Magazine* 77 (March/April 2012). artasiapacific.com/Magazine/77/EscapingGhost.

Zax, David. "A Playboy Model and Nanoscale Printing." *MIT Technology Review*, August 16, 2012. m.technologyreview.com/blog/helloworld/28119/.

Zimmer, Carl. "Nonfiction: Nabokov Theory on Butterfly Evolution Is Vindicated." *New York Times*, January 25, 2011. www.nytimes.com/2011/02/01/science/01butterfly.html?pagewanted=all&_r=0.

Zittrain, Jonathan. *The Future of the Internet and How to Stop It*. New Haven, CT: Yale University Press, 2008.

index